MICROBIAL COMMUNITY ANALYSIS:

THE KEY TO THE DESIGN OF BIOLOGICAL WASTEWATER TREATMENT SYSTEMS

Edited by

T.E. CLOETE and N.Y.O. MUYIMA

Published by the International Association on Water Quality
in its Scientific and Technical Report series.

ISBN 1 900222 02 7
ISSN 1025-0913

British Library Cataloguing in Publication Data

A CIP catalogue record for this book is available from the British Library

This report is published by the International Association on Water Quality,
Duchess House, 20 Masons Yard, Duke Street, St James's, London SW1Y 6BU, England

IAWQ is a company limited by guarantee.
Registered in England No. 1203622. Registered office as above. Registered charity (England) No. 289269.

Typeset in 10½/12 point New Caledonia by Perfect Page Publishing Services, London, England
Index prepared by June Morrison, Helensburgh, Scotland
Cover design by Bernard Fallon Associates, Winchester, England

Printed in Great Britain by the University Press, Cambridge

Contents

List of contributors

Rudolf Amann
Lehrstuhl für Mikrobiologie
Technische Universität München
Arcisstrasse 16
D-80 290 München
Federal Republic of Germany

Ronald M. Atlas
Department of Biology
University of Louisville
Louisville
Kentucky 40292
USA

Thomas E. Cloete
Department of Microbiology and Plant Pathology
University of Pretoria
Pretoria 0002
South Africa

G.A. Ekama
Water Research Group
Department of Civil Engineering
University of Cape Town
Rondebosch 7700
South Africa

M.N.B. Momba
Environmental Biotechnology Laboratory
Department of Microbiology and Plant Pathology
University of Pretoria
Pretoria 0002
South Africa

N.Y. Osée Muyima
Department of Biochemistry and Microbiology
University of Fort Hare
Alice 5700
South Africa

Michael Wagner
Lehrstuhl für Mikrobiologie
Technische Universität München
Arcisstrasse 16
D-80290 München
Federal Republic of Germany

Jiri Wanner
Prague Institute of Chemical Technology
Department of Water Technology and Environmental Engineering
Technika 5
CZ-166 28 Praha 6
Czech Republic

M.C. Wentzel
Water Research Group
Department of Civil Engineering
University of Cape Town
Rondebosch 7700
South Africa

The Water Institute of South Africa
PO Box 6011
Halfway House 1685
South Africa

Introduction

T.E. Cloete

Microorganisms, i.e. algae, bacteria (including cyanobacteria and mycoplasmas), fungi (including lichens and yeasts), protozoa, viroids and viruses, are vital to the function and maintenance of the Earth's ecosystems and biosphere. Estimates of the number of species of microorganisms in the environment range from 5,000,000 to 6,000,000. Less than 5% of these species have so far been described, of which *ca.* 3,500 are bacteria, 90,000 are fungi, 100,000 are protists and 4,000 are viruses. As major contributors to biogeochemical cycles, they perform unique and indispensable activities in the circulation of matter in the world, on which large organisms, including humans depend.

Microorganisms are primary producers and dominate the biogeochemical cycles on our planet. Hence microorganisms have an immense functional diversity and capability and constitute a major genetic resource (information pool) to counteract environmental changes and stresses. Microorganisms transform organic compounds in ways that are exceedingly important for environmental decontamination and remediation. Acting as biochemical incinerators, microorganisms convert pesticides, petroleum and other pollutants in soil and water to harmless products such as carbon dioxide and water. In these ways, microorganisms contribute to waste treatment and to self-purification processes by which aquatic and terrestrial ecosystems are rid of unwanted pollutants.

Biological wastewater treatment systems rely on the interactions and metabolism of microorganisms. Essentially, biological wastewater treatment systems depend on the capacity of the microbial community to recycle elements by way of biogeochemical cycles. The composition of the microbial community is determined by the type of wastewater (nutrient content) and a number of other selective pressures such as mean cell retention time, aerobiosis, anaerobiosis, temperature and other extrinsic factors. The basic principle of 'everything is everywhere, the environment selects' is very applicable to biological wastewater treatment systems. Depending on the physical, chemical and biological selective pressures, we find systems that select for a monoculture in some cases and systems that comprise a complex community of microorganisms in others. This selection for 'natural' microbial populations leads to resistance to changes: these populations can ameliorate stress and are responsible for natural and managed biodegradation of the pollutants. By manipulating the environment, scientists have exploited processes such as nitrogen fixation, nitrification, denitrification, chemolithotrophic carbon dioxide fixation, anaerobic fermentation, photosynthesis, biological phosphorus fixation, sulphate reduction and oxidation, methanogenesis and many other processes in order to treat wastewater. Wastewater treatment processes have evolved by making use of and enhancing Nature's own way of biodegrading organic matter through the recycling of elements.

In the development of biological wastewater treatment systems, scientists have tended to concentrate on the processes rather than on the species involved or on community structure and function. Most of the time, process design is done without fully understanding and/or giving due consideration to the biological processes involved. Consequently they have used simplified approaches, where system function and structure dealt with the minimum of complexity in order to explain the dynamics of the system. Nevertheless, much success has been achieved in this way, indicating the 'forgiving' nature of wastewater, selecting for 'naturally' occurring microbial populations that can biodegrade the pollutant elements.

An understanding of the microbial community of a biological wastewater treatment system would assist in many ways in improving system design and performance. Without the basic knowledge of the microbial community structure and function, it would be impossible to calculate the biodegrading capacity of a system, whether it be a high rate algal pond or an activated sludge system. Microbiology and microbiologists therefore have much to contribute to improve the efficacy of wastewater treatment systems. Management and design of these systems will be successful only if use is made of microorganisms, as tools to detect, assess, restore and maintain information support systems. Hence ecological approaches to effective management of wastewater treatment systems must be based in part on the role that microorganisms play as agents of change and in response to change. Therefore it has become important to introduce some general conceptual issues regarding diversity at the community level. Hypotheses should deal with functional redundancy, species diversity and the stability of communities in biological wastewater treatment systems, and the effect of disturbances on the structure of communities. The more sophisticated the system design, the more difficult and the more important it becomes to understand the community structure and function. Hence mathematical models have been developed to simulate system performance. Detailed microbiological and biochemical information is, however, required for this modelling approach.

Over the years, systems have evolved empirically to what we have today; now we are using community analysis to optimize these systems. The better we understand community structure and function, the better we can manage wastewater treatment systems to control bulking, for example, or to improve biological phosphate removal capacity. Microbial community structure and function has therefore become the driving force in retrofitting, system optimization and system design.

Unfortunately, owing to the lack of suitable technology to determine microbial community structure and function, most wastewater treatment systems lack sophistication. No complete inventory of the microbial species of all wastewater treatment systems exists. The majority of the organisms are viable but non-culturable, making it very difficult to study and manage these systems. Emphasis is therefore placed on those groups of functional importance with reference to system function when it comes to system design and management. In some cases, for example, microbiological indicator organisms are used to monitor and manage system performance. Microbial ecology studies have enabled us to understand and monitor changes in physical, chemical and biological aspects of wastewater systems.

Recent developments have seen the integration of microbial ecology and molecular biology. So-called molecular ecology is rapidly evolving and is giving a new insight into the interrelations that exist between microorganisms and their environment. However, during the past decades of theoretical and empirical research on the causes of patterns in species richness, it has emerged that simple factor explanations are inadequate to account for these patterns. It is believed that the patterns in species diversity are the products of interacting forces that vary in relative importance both in time and space. These interactions are as yet not completely understood.

Our techniques constantly improve for studying microbial community structure and function. This will inevitably lead to improvements in system design, performance and management.

In this Scientific and Technical Report an overview is given of biological wastewater treatment systems and the principles on which they operate. The microbiology of these systems is discussed. The latest techniques for studying community structure and function are reviewed, as well as the principles for system modelling and design.

1. Biological methods for the treatment of wastewaters

N.Y. Osée Muyima, M.N.B. Momba and Thomas E. Cloete

1.1 Introduction

With a rapidly expanding world population, and a highly urbanized and industrialized society, the problems related to the management of wastewater have become of considerable magnitude. It is known that domestic and industrial effluents are principal sources of our natural water pollution load. Nevertheless, the general concern for our natural water resources and the environment is growing. This has resulted in an increasing interest and development of methods and systems by which wastewater can be recycled. The need for technologies for environmentally friendly treatment of wastewater is therefore obvious. Biological processes are a cost-effective and environmentally sound alternative to the chemical treatment of wastewater. This chapter provides an overview of the biological methods for the treatment of wastewaters.

It is important to recall that the present chemical state of the elements in the atmosphere of our planet is a consequence of the metabolism of living organisms. Because all the major elements necessary for life, namely carbon, oxygen, nitrogen, sulphur and phosphorus, are biologically transformed, it is necessary first to summarize the basic cycles of matter on which biological wastewater treatment processes rely. All organisms participate in various steps of the cyclic conversions, but the contribution of microorganisms is particularly important, both quantitatively and qualitatively. The remarkable ability of microorganisms to degrade a vast variety of organic compounds has led to a widely held conviction that "somewhere or other some organism exists which can, under suitable conditions, oxidize any substance which is capable of being oxidized" (Stanier *et al.* 1988) or that "everything is everywhere, the environment selects".

1.1.1 The phosphorus cycle (Figure 1.1)

Of the major nutrient cycles in the environment, phosphorus is the simplest. Phosphorus is found either as inorganic (phosphate ion) or organic phosphate. Phosphorus requirements for living organisms are met only by the uptake of phosphate ions. Organic phosphate is then synthesized within the cell and, on the death of the organism, phosphate is rapidly released by hydrolysis. Phosphate is a limiting factor for the growth of many organisms because much of our planet's supply of phosphates occurs as insoluble complexes (e.g. calcium, iron or aluminium salts). Freshwater, for instance, often contains phosphate ions in trace amounts.

Soluble phosphates are constantly transferred from terrestrial environments to the sea as a consequence of leaching, a transfer that is largely unilateral. For instance, losses of phosphate during decomposition have been reported in several mangrove forests (Wiebe 1989). The availability of phosphate for terrestrial organisms therefore depends on the continued solubilization of insoluble phosphate deposits, a process in which microorganisms play a key role.

1.1.2 The oxygen cycle (Figure 1.2)

There are two principal reservoirs of oxygen: gaseous O_2 and water. Gaseous oxygen is generated from water almost exclusively by oxygenic photosynthesis performed by higher plants, algae and cyanobacteria. The reverse conversion of oxygen to molecular form is mediated by all organisms that perform aerobic respiration, of which microorganisms are quantitatively the most important. Combustion of fossil fuels has become an increasingly important pathway by which atmospheric oxygen is utilized.

CO_2 is the second main product of respiration. However, it is more a critically important intermediate in the cyclic interconversion of carbon-containing compounds than a reservoir of oxygen.

1.1.3 The carbon cycle (Figure 1.3)

The concentration of CO_2 in the atmosphere, like that of O_2, is largely determined by the competing processes of photosynthesis and respiration, although other processes can contribute. However, compared with the amount of oxygen in the atmosphere (about 21% by volume), the concentration of CO_2 is very low (about 0.03% by volume). The concentration of CO_2 is therefore the limiting factor for plant growth under favourable environmental conditions of light intensity and temperature.

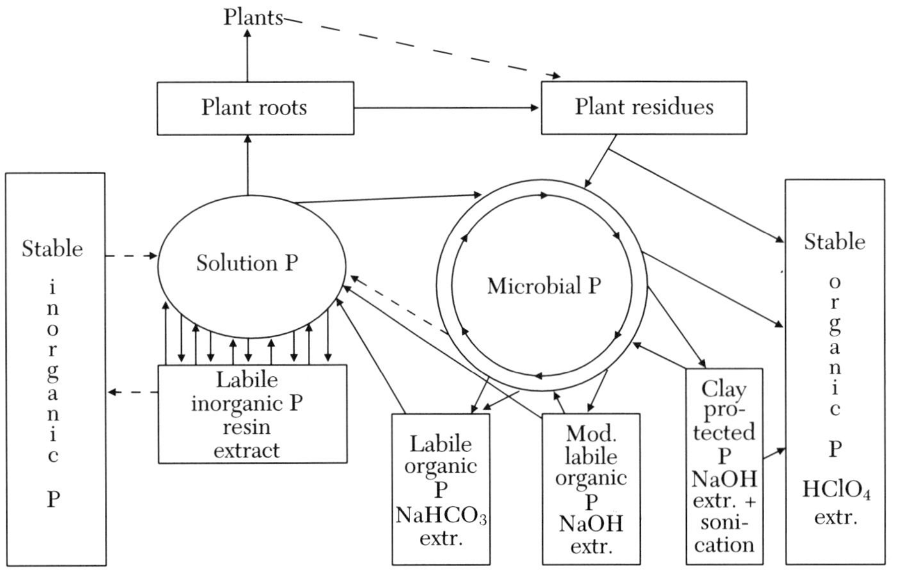

Figure 1.1. The phosphorus cycle.

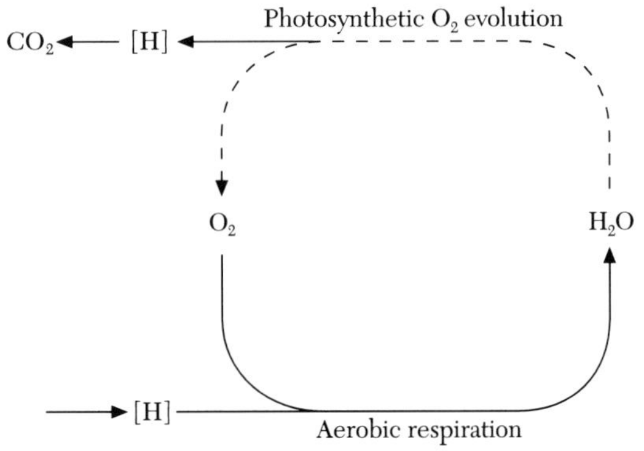

Figure 1.2. The oxygen cycle.

When CO_2 is dissolved in slightly alkaline water, bicarbonate and carbonate ions are produced. Bicarbonate serves as the reservoir of carbon for photosynthesis in aquatic environments. The bicarbonate concentration of oceans trap a large fraction of the CO_2 produced on land, keeping its atmospheric concentration at a relatively low and constant level.

The biological conversion of organic carbon to CO_2 with the concomitant reduction of molecular oxygen involves the combined metabolic activity of several different kinds of microorganism. Many aerobic bacteria, as well as fungi, perform the complete oxidation of organic substances. In the oceans, CO_2 from aerobic respiration produces carbonate ions, which combine with dissolved calcium ions and are finally sequestered as calcium carbonate. Sequestration of carbon can also take place as organic deposits. Indeed, much carbon has been sequestered from the biosphere in the form of peat and coal.

In anaerobic environments, organic compounds are decomposed initially by fermentation; the organic end products of fermentation are then further oxidized by anaerobic bacteria, in the presence of suitable inorganic hydrogen acceptors (such as nitrate, sulphate or CO_2).

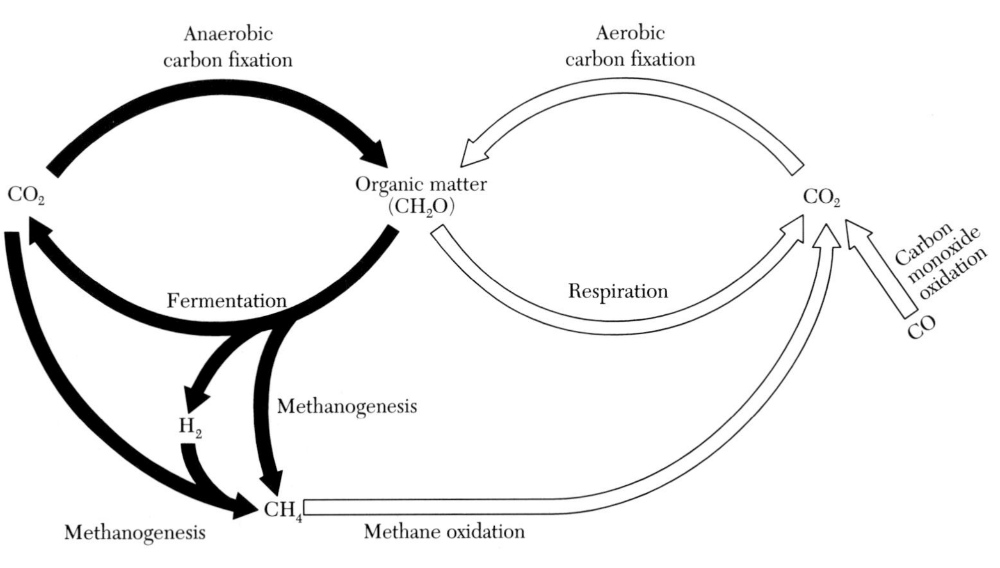

Figure 1.3. The carbon cycle.

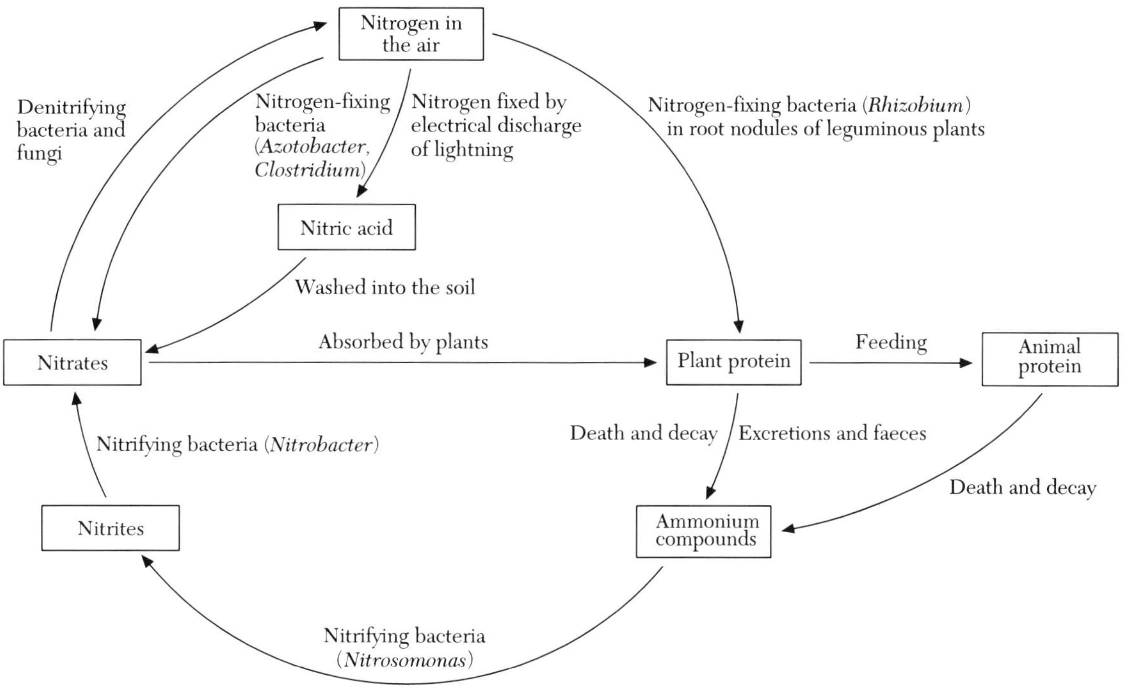

Figure 1.4. The nitrogen cycle.

1.1.4 The nitrogen cycle (Figure 1.4)

Although molecular nitrogen is abundant, constituting about 79% of the Earth's atmosphere, it is chemically inert and therefore not a suitable source of the element for most living forms. Most living organisms depend on a source of combined nitrogen (i.e. ammonia, nitrate and organic compounds) or fixed nitrogen for their nutrition. Combined nitrogen is scarce in soil and water. It often constitutes the limiting factor for the development of living organisms. The cyclic transformation of nitrogenous compounds is therefore of most importance in supplying required forms of nitrogen to the various nutritional classes of organism in the biosphere.

Nitrogen fixation is largely a biological process carried out by bacteria. These include non-symbiotic bacteria, e.g. *Azotobacter, Mycobacterium* and *Thiobacillus*; methane oxidizers as free-living aerobic bacteria, e.g. *Clostridium, Klebsiella, Bacillus, Desulfovibrio* and *Desulfotomaculum*; methanogenic bacteria as free-living anaerobic bacteria; and symbiotic bacteria, e.g. *Rhizobium, Azospirillum, Frankia* (aerobic) and *Citrobacter* (anaerobic). The symbiotic fixers are quantitatively more important. Because of the critical agronomic importance of fixed nitrogen, biological nitrogen fixation has become an intensive subject of investigation (Wiebe 1989).

The immediate product of nitrogen fixation, ammonia, enters the environment and is oxidized to nitrite and then nitrate (nitrification); in this form plants and many soil bacteria assimilate nitrogen. The conversion of ammonia to nitrate is brought about in nature by two highly specialized groups of obligatory aerobic chemoautotrophic bacteria in which, respectively, nitrification occurs in two steps. In the first step, ammonia is oxidized to nitrite; in the second, nitrite is oxidized to nitrate. Nitrates are very soluble compounds and are therefore easily leached from the soil and transported by water. Hence a certain amount of combined nitrogen is constantly removed from the continents and is carried down to the oceans. Moreover, many aerobic bacteria can use nitrate in place of oxygen as a final electron acceptor under anaerobic conditions: this opposite process is termed denitrification. The reduction of nitrate to nitrite can be pursued up to the nitrous oxide gas level and subsequently to nitrogen gas. Denitrification has a major ecological effect because it depletes the soil of an essential nutrient for plants, thereby decreasing agricultural productivity.

1.1.5 The sulphur cycle (Figure 1.5)

The turnover of sulphur compounds is referred to as the sulphur cycle. Sulphur, an essential constituent of living matter, is the tenth most abundant element in the Earth's crust. It is available to living organisms principally in the form of soluble sulphates or reduced organic sulphur compounds.

In addition to the biological sulphur cycle, important non-biological transformations of gaseous forms of sulphur occur in the Earth's atmosphere, for example SO_2 is contributed by the burning of fossil fuels, H_2S and SO_2 arise from the Earth's volcanic activity. In the atmosphere, H_2S is rapidly oxidized by oxygen (i.e. atomic, molecular) or ozone to SO_2 which

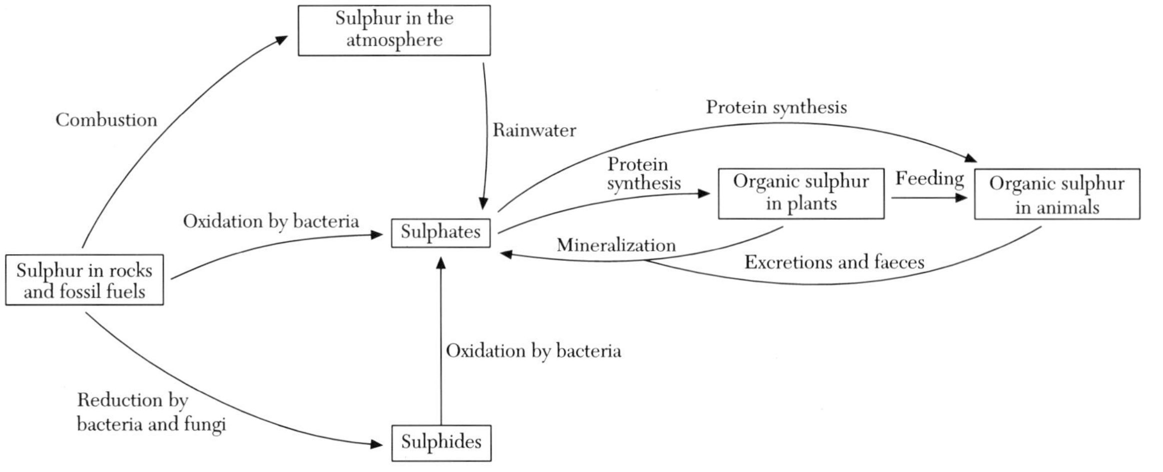

Figure 1.5. The sulphur cycle.

can dissolve in water to form sulphurous acid or be oxidized slowly to SO_3. When dissolved in water, SO_3 becomes sulphuric acid. Most of the sulphuric acid returns along with unoxidized sulphurous acid to the Earth's surface in acid form, where it causes considerable harm (damage to stone structures and sculptures, and killing of the flora and fauna of lakes).

Sulphur as sulphate is in abundant supply in sea water. The assimilation of sulphate by plants and microorganisms resembles the assimilation of nitrate. The sulphur atom of sulphate must become reduced to be incorporated into organic compounds, because in living organisms sulphur occurs mostly in a reduced form as –SH or –S–S– groups.

When sulphur-containing organic compounds are mineralized, sulphur is liberated in a reduced inorganic form as H_2S. The utilization of sulphate for the synthesis of sulphur-containing cell constituents and the subsequent decomposition of these compounds results in an overall reduction of sulphite to H_2S. H_2S is also formed more directly from sulphate through the activity of the sulphate-reducing bacteria. This activity is particularly apparent in the mud at the bottom of ponds and streams, in bogs and along the seashore. The H_2S that is produced in the biosphere as a result of the decomposition of sulphur-containing compounds, of the reduction of sulphate and of volcanic activity is in turn converted to sulphate by photosynthetic and chemoautotrophic bacteria. Only a small part of it becomes sequestered in the form of insoluble sulphides or, after spontaneous oxidation with oxygen, as elemental sulphur.

By manipulating the external environment through process design, it is possible to create favourable conditions for the cycling of CO_2, O_2^{2-}, SO_4^{2-}, nitrogen and phosphorus in wastewater treatment systems. These applications will be discussed in the context of different types of biological wastewater treatment system.

1.2 The activated sludge process

Activated sludge is at present the most widely used biological treatment process for both domestic and industrial wastewaters. The activated sludge process refers to a continuous or semi-continuous (fill-and-draw) aerobic method for biological wastewater treatment, including carbonaceous oxidation and nitrification. The process relies on a dense microbial population's being mixed in suspension with the wastewater under aerobic conditions. In the presence of adequate nutrients and oxygen a high rate of microbial growth and respiration is achieved. This results in the utilization of the organic matter present in the production of oxidized end products such as CO_2, NO_3^-, SO_4^{2-} and PO_4^{3-}, and/or the biosynthesis of more microorganisms. Activated sludge treatment removes from the wastewater the biodegradable organics as well as the unsettleable suspended solids and other constituents, which can be adsorbed on, or entrapped by, the activated sludge floc.

According to Grady and Lim (1980), the expression 'activated sludge' alludes to a slurry of microorganisms that remove organic compounds from wastewater; these microorganisms are themselves removed by sedimentation under aerobic conditions. In an activated sludge system, soluble biodegradable organic compounds are degraded by bacteria in an aerated basin, and biomass is carried over with the influent into a clarifier where solids are allowed to settle and concentrate; they are then removed. Part of the settled sludge is drawn off as waste; the rest is recycled to the aeration basin to maintain a high concentration of bacteria (Eckenfelder *et al.* 1985). In the conventional activated process, which is not designed or operated to achieve the biological removal of excess phosphate, the bacteria use phosphate only in quantities that satisfy their basic metabolic requirements. Because of the usual nutrient imbalance in sewage, only a limited quantity of the feed phosphate will be removed in such plants (Pitman 1984).

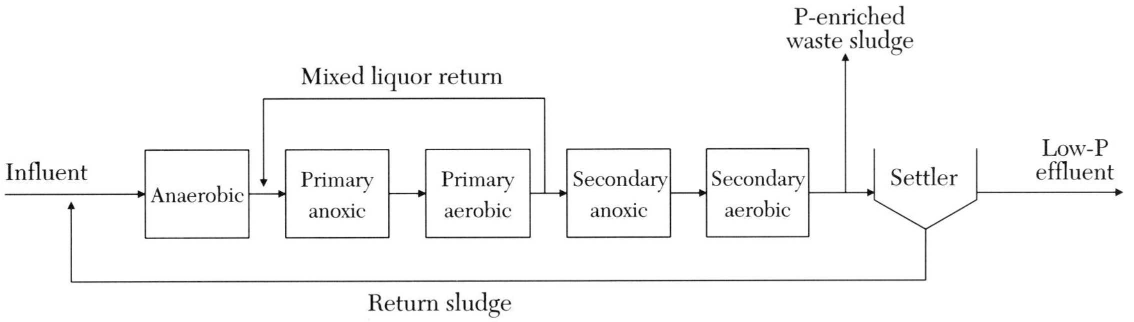

Figure 1.6. Diagram of the five-stage Phoredox or Modified Bardenpho process for carbon, nitrogen and phosphate removal (Toerien et al. 1990).

The activated sludge process has undergone various modifications able to meet most wastewater treatment needs (Toerien *et al.* 1990). Briefly, these changes relate to the flow regime in the reactor, the size, number and configuration of the reactors, recycled flow, influent flow and others incorporated either emphatically or current inadvertently or unavoidably. Whereas the response of the organisms is in accordance with their nature, that of the process is governed by both the organism characteristics and the physical features that define the process (Ekama and Marais 1984). Although the process has undergone various adjustments, the basic principles stay the same: the removal of soluble biodegradable organic compounds from wastewater by a flocculant slurry of microorganisms, the removal of microorganisms by sedimentation, the recycling of a small amount of return sludge from the clarifier underflow to the reactor (primary aerobic zone) and the dependence of the systems' high performance on the mean cell resistance time (MCRT) of microorganisms (Grady and Lim 1980).

The activated sludge process has been developed primarily for the removal of carbon, nitrogen and phosphate. The basic process for the simultaneous biological removal of phosphates and nitrogen was proposed by Barnard in 1976 and is known as the Phoredox process in South Africa and the Bardenpho or Modified Bardenpho in the United States (Figure 1.6). The Bardenpho activated sludge process is a four-stage process designed mainly for removing nitrogen, whereas the Phoredox activated sludge process is a five-stage process designed to remove nitrogen and phosphate. Nevertheless, the two systems consist of a sequence of primary anoxic, primary aerated, secondary anoxic and secondary aerated basin, followed by a clarifier. In the Phoredox activated sludge process, the incorporation of an anaerobic zone at some point in the process allows the release of phosphate; after being released from the biomass in the anaerobic zone, phosphate is reincorporated into the biomass during aerobiosis, together with part or all of the influent phosphate (Gerber *et al.* 1986). It has been indicated that the release of phosphate is a basic requirement for successful uptake of phosphate. Anaerobic–aerobic or anoxic–oxic activated sludge systems are characterized by their capacity for efficient removal of phosphate from wastewater (Barnard 1976). In these systems, sludge is returned from the clarifier to re-enter the anaerobic zone with the influent. Mixed liquor is returned from the primary aerobic zone to the primary anoxic zone, and so on. The application of this process in South Africa has led to the removal of phosphate in the final effluent of activated sludge plants to levels between 0.2 and 0.8 mg/l, together with the removal of 80–90% of the nitrogen (Barnard 1976).

Activated sludge systems modified for the enhanced removal of phosphate during effluent treatment are now employed widely in at least eight countries, i.e. Australia, Brazil, Canada, France, New Zealand, South Africa, the United States and Zimbabwe (Toerien *et al.* 1990). Wastewater treatment makes its contribution by controlling wastage and safeguarding available resources against pollution. Effective removal of the pollutants from wastewater before discharge is of prime importance and the production of high-quality effluents can have an economic advantage by supplementing existing water resources (Slim 1987). The primary objective of the activated sludge process is the removal of soluble biodegradable compounds. The processes involved in nutrient removal must be not only technically possible but also practicable and must be able to meet the effluent criteria in an economical manner. The activated sludge process has been renowned for producing an effluent of high quality at reasonable cost. It is capable of achieving equal reductions in soluble substrate in reactors of much smaller volume while producing an effluent relatively free of suspended solids (Grady and Lim 1980). The activated sludge process removes certain priority pollutants with an efficiency of 95% or more (Eckenfelder *et al.* 1985). Moreover, the activated sludge process is controllable, because through adjustment of the amount of sludge wasted, the operator is able to regulate the MCRT to obtain the desired effluent quality. Activated sludge reactors are relatively resistant to shock loads and can achieve acceptable effluent in spite of dynamic inputs (Grady and Lim 1980), although they do perform better under more stable conditions.

However, like most processes, the activated sludge process also presents some problems. The main problem is a direct result of its controllability. It requires relatively sophisticated operation to achieve the desired results. The decision to install such a system requires considerable commitment on the part of an industry or municipality (Grady and Lim 1980). To understand the process better, a description of the different stages is necessary.

1.2.1 Anaerobic zone

The anaerobic zone is considered to be one in which both dissolved oxygen and oxidized nitrogen (nitrate or nitrite) are absent (Barnard 1976; Buchan 1984). In this zone, sludge from the clarifier flows jointly with the influent wastewater. The anaerobic zone is essential for the removal of phosphate, because the bacteria in the activated sludge passing through this zone are preconditioned to take up excess phosphate under aerobic conditions. The release of a certain quantity of phosphate from the biomass into the solution indicates that the bacteria have been suitably conditioned (Pitman 1984). In addition to the anaerobic conditions, the retention time (about 1 h) of the influent wastewater is of extreme importance. Conditions that are likely to affect either, such as return flows with high dissolved oxygen or nitrate concentration, must be avoided (Keay 1984; Tam *et al.* 1992). Nitrates and dissolved oxygen discharged into this zone must be zero or as near to zero as possible at all times.

The presence of nitrate in an anaerobic zone has been reported to be a handicap to the phosphate-removing potential of the activated sludge system (Barnard 1976; Nicholls and Osborn 1979; Marais *et al.* 1983). High concentrations of nitrate present in the anaerobic zone resulted in poor removal of phosphate during aerobiosis. With an incomplete degree of nitrification, the phosphate removal was 55% (Mulder and Rensink 1987). However, too many factors were involved for a good correlation between the phosphate removal and nitrification (Mulder and Rensink 1987). Presumably, competition for substrate between *Acinetobacter* and denitrifying organisms is involved. However, insufficient denitrification is unacceptable for the requirements of effluent standards (Mulder and Rensink 1987). Moreover, in the presence of nitrate, the redox potential is too high to produce lower fatty acids for the release of phosphate. The availability of lower fatty acids under these circumstances seemed to be too low for good release of phosphate. The use of unsettled influent and the presence of sludge from the sludge treatment in the primary clarifiers, probably producing lower fatty acids, had a positive effect on the phosphate removal (Mulder and Rensink 1987). The influent nitrate levels should be low to ensure that nitrates returned with the underflow from the final clarifier do not negatively affect the performance of the initial oxygen-limiting zone, and also to ensure the achievement of a good release of phosphate. The degree of nitrate feedback that can be tolerated depends on the strength of the sewage feed to the anaerobic zone and in particular its readily biodegradable COD concentration (Pitman 1984). It has been indicated that the bio-polyphosphate bacteria were generally not capable of nitrate reduction. However, Lötter (1985) detected a number of *Acinetobacter* isolates able to reduce nitrate to nitrogen gas. This finding could explain the limited uptake of phosphate in the anoxic zone.

It has been proved that in the anaerobic zone the microorganisms normally living in soil and water and that are capable of fermentation (species of *Aeromonas, Citrobacter, Klebsiella, Pasteurella, Proteus and Serratia*) accumulate and produce organic compounds such as lactic acid, succinic acid, propionic acid, butyric acid, acetic acid and ethanol during fermentation. These organic compounds serve as electron donor and acceptor, but cannot be utilized under anaerobic conditions. They will be consumed only in the anoxic and aerobic zones. Therefore it seems as though the anaerobic zone provides substances for the proliferation of aerobic phosphate-accumulating bacteria (Fuhs and Chen 1975; Buchan 1984).

Certain carbon substrates induce phosphate release, whereas others stimulate phosphate uptake. The addition of acetate to the medium and a lowering of the pH under anaerobic conditions (Fuhs and Chen 1975), as well as acetate treatment and phosphate starvation, resulted in phosphate release (Barnard 1976; Buchan 1983). Phosphate release by the addition of acetate has also been observed by Comeau *et al.* (1986) and Murphy and Lötter (1986). Electron microscopic examination showed the disappearance of intracellular polyphosphate granules during acetate treatment; the release of phosphate into the external medium was therefore based on the presence of acetate (Murphy and Lötter 1986). Lötter's (1985) observations indicated that the addition of acetate caused the extracellular orthophosphate level to rise to 55 mg/l. The release of orthophosphate, as triggered by the addition of acetate, could well be a side effect of acetate's ability to dissipate the proton motive force that controls the movement of metabolites across cell walls. Once the proton motive force has been weakened, orthophosphate readily diffuses to the external medium until equilibrium has been reached. Organics such as acetate and propionate induce phosphate release from activated sludge. The release of phosphate occurs immediately on exposure to acetate or propionate, irrespective of whether the sludge is partly or fully reconstituted with respect to accumulated phosphate (Gerber *et al.* 1987). Whereas Lötter (1985) contested the effect of succinate on phosphate release, Gerber *et al.* (1986) maintained that succinate plus other short-chain carbon compounds such as citrate, glucose, ethanol, methanol and 2,3-butanodiol induced the release of phosphate after the onset of anaerobiosis. The most favourable net removal of phosphate from solution is obtainable by the use of acetate, butyrate, propionate and lactate, but formate stimulates good release of phosphate (Gerber *et al.* 1986). Nevertheless, the removal of phosphate, stimulated by short-chain carbon compounds, depends on the substrate concentration. Bosch (1992) indicated that an environment lower in nutrient (i.e. 200 mg/l sodium acetate) resulted in less removal of phosphate (9 mg/l) from the mixed liquor medium, whereas a higher concentration of nutrient (i.e. 5 mg/l sodium acetate) led to a greater removal of phosphate (20 mg/l).

Fuhs and Chen (1975) were the first workers to relate phosphate release to the subsequent phosphate uptake. According to these workers, the presence of the *Acinetobacter* is as a result of the anaerobic–aerobic sequences' acting as a selective pressure. The principal function of the anaerobic zone is to establish a facultatively anaerobic microbial community as indicated by the Embden–Meyerhof fermentation pattern. During anaerobiosis this bacterial community tends to produce compounds such as ethanol, acetate and succinate, which serve as carbon sources for *Acinetobacter*. This supports the hypothesis of those workers who consider that the anaerobic zone is an essential zone for the formation of substrates that could ensure the proliferation of *Acinetobacter* (Fuhs and Chen 1975; Deinema *et al.* 1980; Buchan 1983; Lötter 1985). It is likely that the absence of anaerobiosis leads to an obligatorily aerobic group developing, in which case the mentioned products would not be formed. Contrary to Shapiro (1967) and Shapiro *et al.* (1967) Fuhs and Chen (1975) postulated that phosphate release is an indication that fermentation is occurring in the anaerobic zone, but is not necessarily a prerequisite for phosphate uptake. Anaerobiosis is not itself necessary to trigger the release of phosphate from the bacteria into the supernatant, so the accumulation of CO_2 and the lowering of the pH that results from anaerobiosis might be important factors. This was also observed by Shapiro (1967) and Shapiro *et al.* (1967). Gerber *et al.* (1986) demonstrated that anaerobiosis provided the environment in which the stated compounds are converted to products that trigger phosphate release rather than itself bringing about release.

Although these workers have indicated the link between the release and uptake of phosphate, many other investigators have attempted to illustrate other factors able to affect both phenomena in an anaerobic–aerobic sequence. Harold (1966) found that the release and uptake of phosphate were linked to some stress situations in anaerobic–aerobic cycles. He suggested that the temporary limitation of sulphur or nitrogen can result in the accumulation of polyphosphate granules within the cells of certain bacteria. It was also demonstrated that the temporary restriction of the availability of oxygen and nitrate led to a storage of polyphosphate. Nicholls and Osborn (1979), by using a pilot plant with an anaerobic–anoxic–aerobic basins, confirmed that bacteria relieved from stress conditions very rapidly take up phosphate in an overplus reaction immediately on entering the aerobic zone, where stress conditions are non-existent and phosphate together with an abundant source of energy is available. In contrast, Ohtake *et al.* (1985) indicated that *Acinetobacter* cells released phosphate under anaerobic conditions, but the rapid and extensive uptake was not induced when it was exposed to aerobic conditions; thus anaerobiosis stress is still insufficient for stimulating the phosphate uptake ability of cells. This led to the conclusion that phosphate release under anaerobic conditions was due to the depletion of energy required by maintaining intracellular levels of phosphate, but not to lysis of the cells. According to Shapiro (1967), the release is apparently triggered by the lack of oxygen and/or low redox potential (such as −150 mV). However, Shapiro *et al.* (1967) indicated that the release of phosphate is more closely related to redox conditions than the dissolved oxygen concentration and that the phenomenon was reversible. Moreover, the phosphate seemed to be released from the acid-soluble fraction of cells and to a minimal degree from RNA and DNA protein fractions.

1.2.2 Primary anoxic zone

In this context, anoxic refers to the presence of nitrates and the absence of dissolved oxygen (Buchan 1984; Pitman 1984; Streichan *et al.* 1990). The primary anoxic zone is the main denitrification reactor in the process. It is fed by the effluent from the anaerobic zone and by the mixed liquor recycled from the aerobic zone. The mixed liquor recycling rate from the aerobic zone to the anoxic zone is a variable that can be controlled. The introduction of supplementary dissolved oxygen into the anoxic zone must be avoided (Keay 1984). The absence of oxygen and the presence of nitrate or nitrite leads to the enrichment of denitrifying bacteria, which reduce nitrate or nitrite to molecular nitrogen. Soluble and colloidal biodegradable matter are readily removed in the primary anoxic zone.

It has been reported that phosphate release occurred under anoxic conditions if suitable carbon substrates were added. Gerber *et al.* (1986) revealed that certain lower fatty acids or their salts such as acetate, formate and propionate were capable of inducing phosphate release from sludge under anoxic conditions. Tam *et al.* (1992) pointed out that a significant increase in wastewater orthophosphate content was found in both glucose and sodium acetate reactors, even when the nitrate concentration in these reactors was still around 10 mg/l after 1 h in the anoxic stage. This suggested that phosphate could be released during anoxic conditions. According to Tam and co-workers (1992), the amount of phosphate released during anoxic–anaerobic stages is affected not only by the substrate, as postulated by Gerber *et al.* (1986), but also by the dosage of the substrate concentrations of soluble, readily biodegradable carbon substrates.

1.2.3 Primary aerobic zone

The main function of this zone is to oxidize organic material in the sewage, to oxidize ammonia to nitrite and then to nitrate and to provide an environment in which the biomass can take up all the phosphate released in the anaerobic zone, plus all the phosphate that enters the process in the feed sewage. The aerobic zone must therefore be sized for complete nitrification (Keay 1984; Pitman 1984). Chemoautotrophs are responsible for this oxidation; ammonia is oxidized to nitrite by *Nitrosomonas, Nitrosospira* and *Nitrosolobus* spp., whereas nitrite is oxidized to nitrate by *Nitrobacter, Nitrospira* and *Nitrococcus* spp. (Buchan 1984).

Various factors can affect the rate of uptake of phosphate by bacteria; ideally the uptake of phosphate should be complete at the end of the aerobic zone, so that only trace quantities (i.e. less than 0.2 mg P/l) can be detected in the solution at this point. The aeration rate seems to be the principal operational determinant of the efficiency of phosphate removal. It should therefore be adequate to promote the rapid uptake of released plus feed phosphate, to ensure the oxidation of carbon compounds and ammonia and to suppress the growth of filamentous microorganisms that produce poorly settling sludges (Pitman 1984).

The efficiency of phosphate removal seems to be dependent on the intensity of aeration. This has been reported by many researchers (Levin and Shapiro 1965; Shapiro 1967; Carberry and Tenny 1973). Levin and Shapiro (1965) demonstrated the storage of phosphate under aerobic conditions. Their experiments showed that, at higher aeration rates, maximum uptake of phosphate was achieved within 2 h. Moreover, the mixed liquor supplied with pure oxygen took up a great deal more orthophosphate than did the mixed liquor supplied with an equal quantity of oxygen in air. Small changes in aeration rate showed differences in orthophosphate uptake of the sludge organisms. Shapiro (1967), in contrast, indicated that although phosphate is released under anaerobic conditions, the situation could be reversed in subsequent aeration. According to Barnard (1976), enhanced removal of biological phosphate could be achieved in a simple activated sludge plant not designed for nitrogen removal by adding an anaerobic zone to the head of the plant (Figure 1.6). In this case, nitrification in the aerobic zone would have to be minimized and then NO_3-N is not discharged into the anaerobic zone via the return sludge. Carberry and Tenny (1973) indicated that 315 ml of air/min per litre was approximately the optimum aeration rate to induce phosphate removal during secondary aeration treatment. This observation led to the conclusion that the aeration rate seemed to be the principal operational determinant of the efficiency of phosphate removal. The findings of Wentzel *et al.* (1985) indicated that the excessive phosphate uptake in the aerobic stage was associated directly with the degree of phosphate release during the previous anaerobic phase; more phosphate release leads to more phosphate uptake.

Though anaerobiosis and aerobiosis play a role in biological phosphate removal in wastewater treatment plants, other parameters affecting the rate of release and uptake include the presence of certain substrates as previously noted. Enhanced removal of phosphate is dependent on the presence of the readily biodegradable compounds, especially volatile fatty acids, produced by fermentative bacteria from organic compounds in the influent. Volatile fatty acids are removed in the anaerobic phase and polymerized at the expense of energy obtained from the breakdown of polyphosphate. There is a linear relation between phosphate release and uptake (Wentzel *et al.* 1985). Efficient uptake of phosphate requires high concentrations of fermentable substrates or volatile fatty acids in the influent to the anaerobic zone. Therefore the addition of volatile fatty acids to the anaerobic zone has been used as a means of enhancing biological removal (Buchan 1983; Mulder and Rensink 1987).

It has been suggested that poly-β-hydroxybutyric acid (PHB) forms an important key intervening in the mechanism of phosphate uptake and release (Nicholls and Osborn 1979; Deinema *et al.* 1980; Lawson and Tonhazy 1980; Comeau *et al.* 1986). This was illustrated by *Acinetobacter calcoaceticus* strains that accumulated PHB in activated sludge systems designed for enhanced phosphate removal, which exhibited an increased rate of phosphate uptake under increased aeration (Lawson and Tonhazy 1980). Phosphate uptake by the biomass was observed when PHB reserves were consumed (Comeau *et al.* 1986). The presence of PHB in bio-polyphosphate bacteria should help them to grow and rebuild their own polyphosphate by taking up soluble phosphate from solution (Comeau *et al.* 1986). At that time, PHB and polyphosphate would play a mutually interdependent role to assist aerobic bacteria to survive through an anaerobic period (Nicholls and Osborn 1979).

1.2.4 Secondary anoxic zone

Further denitrification takes place in this zone. The main function of the secondary anoxic zone is the removal of excess nitrates not removed in the primary anoxic zone. Because of the very slow denitrification rate in this zone, the quantity of nitrate removed by it is small. The retention time in the anoxic stage is relatively long because of the lower COD.

1.2.5 Secondary aerobic zone and clarifier

The mixed liquor continues its circuit from the secondary anoxic zone to the secondary aerobic zone. The latter is required to increase the dissolved oxygen to a level of between 2 and 4 mg/l in the mixed liquor before it enters the clarifier (Barnard 1976), and to refine the final effluent by the removal of additional phosphate and the oxidation of residual ammonia. Mixed liquor must be aerated for a period at least 1 h before it passes into the clarifier. Aeration should be satisfactory to promote phosphate uptake and to maintain good aerobic conditions. However, excess aeration should be prevented as it will encourage the conversion of organically bound nitrogen to nitrate and also cause the slow aerobic release of phosphate from the solid (Keay 1984; Pitman 1984). This zone prevents anaerobic conditions from developing in the clarifier and also prevents phosphate release before clarification. Residual ammonia in the mixed liquor will continue to be nitrified and if phosphate has not been completely removed in the previous zone it will continue to be removed in this zone (Buchan 1984).

The aim of the clarifier is to produce a clear effluent free of suspended solids, and a thickened sludge for recycling to the inlet of the process. The quality of the underflow sludge should be such that nitrate is not recycled to the anaerobic zone (Ekama *et al.* 1984).

1.2.6 Principles of the activated sludge process

1.2.6.1 Concept of aerobic treatment

The activated sludge process depends on the physiology of heterotrophic organisms that, in the presence of oxygen, utilize the organic substances present in the wastewater as a carbon source for cell synthesis and as an energy source. When wastewater mixes with the microorganisms of activated sludge in the presence of dissolved oxygen, two phenomena responsible for the initial removal of biodegradable organic substances (expressed as BOD (biochemical oxygen demand)) take place. The suspended and colloidal solids and, to a certain extent, the soluble organic substances in the wastewater are absorbed on the surface of activated sludge flocs. Concurrently, intense biological activity converts part of the wastewater organics into a reserve food inside the microbial cells in the sludge. Further wastewater organic matter is progressively removed by continued aeration. The latter rate of removal depends on the remaining BOD and the concentration of the activated sludge.

1.2.6.2 Nutrient requirements

With regard to nutrient requirements, nitrogen and phosphorus are the most important nutrients in the activated sludge process for two reasons:

(1) The lack of these nutrients might significantly affect the growth of activated sludge/biofilm microorganisms. It can also result in a selection of undesirable filamentous microorganisms.
(2) Elevated concentrations of nitrogen and phosphorus in effluents from biological wastewater treatment plants contribute to the eutrophication of receiving waters.

Phosphorus is directly involved in biosynthesis, whereas nitrogen is involved in the energy transfer system microorganisms. Other mineral nutrients such as Mg, K, Ca, Fe, Mn, Cu and Co are needed in very small amounts for cell metabolism.

1.2.6.3 Flocculation of microorganisms

The floc or clump of microorganisms is the basic operational unit of activated sludge. Although the exact nature of the flocculating material is not well known, it seems almost bacterially originated. Good flocculant growth is necessary for the successful operation of the process because suspended, colloidal and ionic matter in the wastewater are removed by adsorption and agglomeration in the aeration tank. In addition, good flocculant growth is important in the sedimentation tank for subsequent rapid and efficient separation of sludge from the treated effluent. The adsorption capacity of the floc depends on the availability of suitable cell surfaces. Once all the adsorption sites are occupied, the floc has a very limited capacity for adsorbing further material until it has metabolized that which it has already adsorbed. Breakdown and assimilation of the agglomerated material, meaning stabilization, indeed proceeds more slowly. The removal process depends on continuous re-inoculation with recycled settled sludge, so the system will select only floc-forming organisms that settle rapidly in the sedimentation tank. Thus the process is microbially self-regulating with the required selected flocs recirculated.

1.2.6.4 Substrate transfer

The stabilization of the organic substrate by activated sludge microorganisms takes place inside the bacterial cells. The transport of substrates into the cell and that of metabolic products out of the cell is therefore important. Among the major transport phenomena involved in activated sludge treatment are the following: transport of gaseous oxygen and substrate into solution, transport of dissolved oxygen and substrate to the vicinity of a metabolizing organism, transport of dissolved oxygen and substrate into cell, and transport of metabolic products out of the cell. All the above transfer mechanisms are mostly diffusion in nature. Thus they are controlled by the respective diffusion coefficient values. However, active transport phenomena do occur, which would explain the relatively high transport rates observed in some cases.

1.2.6.5 Nitrification and denitrification

Owing to microbial processes in sewers, nitrogen is present in raw wastewaters mostly in a reduced form, i.e. as ammonia nitrogen and organic nitrogen (amino acids, proteins, nitrogen heterocyclic compounds). During carbonaceous oxidation of wastewater, many forms of organic nitrogen are converted to ammonia nitrogen. Ammonia, the reduced form of nitrogen, is in turn oxidized by autotrophic nitrifying bacteria to nitrate via nitrite, through a process known as nitrification. The heterotrophic reduction of nitrite and nitrate to molecular nitrogen leads to a substantial elimination of nitrogen from wastewater.

The nitrification process requires a substantial amount of oxygen and has an effect on the decrease in wastewater alkalinity. Such a decrease in wastewater alkalinity might cause a decrease in its pH when the alkalinity of the wastewater is low or its ammonia content relatively high. Nitrification is possible at a relatively broad range of pH values; however, the optimal results are reported mostly between pH 7.5 and 8.5. The temperature of mixed liquor does affect the rate of nitrification. Heavy metals in the ionic form are inhibitors of the nitrification process. However, at optimal pH values for nitrification, these metals are relatively insoluble and therefore do not affect the process at concentrations of 10–20 mg/l. Dissolved oxygen contents might have an influence on the nitrification process. Effective nitrification has been reported in systems with a dissolved oxygen contents of only 0.5 mg/l. According to Ekama and Marais (1984) the oxygen requirement for nitrification of domestic wastewater, for instance, is *ca.* 25–35% of the total oxygen requirement. A large part of the oxygen demand for nitrification can be recovered when nitrification is followed by denitrification.

To decrease the overall cost of nitrification and to decrease the concentration of nitrate being discharged in the final effluent, modern activated sludge plants that are able to nitrify wastewaters often incorporate a facility for denitrification. This is a biological process that occurs under conditions of low dissolved oxygen, where oxidized forms of nitrogen (nitrite or nitrate) are reduced to molecular nitrogen. A wide range of heterotrophic bacteria can accomplish denitrification under anoxic conditions. The denitrification stage is of extreme importance in full-scale nutrient removal activated sludge plants. The denitrification process is slower below pH 6.0 and above pH 8.7. The highest rates of denitrification occur within the narrow pH range of 7.0–7.5. Denitrification, although temperature-dependent, can take place at temperatures as low as 5 °C.

1.2.6.6 Phosphorus removal

The bulk of phosphorus in wastewaters comes from detergents, both domestic and industrial. Phosphorus is present in sewage in both inorganic and organic forms. Phosphorus present in raw wastewaters is often converted to orthophosphate. In this more assimilable form, phosphorus enters the biochemical reaction system.

Phosphorus removal from wastewater can be achieved through chemical precipitation and coagulation, through biological treatment, or through a combination of both. At present the most common biological wastewater treatment method is the aerobic activated sludge process. According to Pauli (1994), enhanced biological removal of phosphorus is based on the enrichment of activated sludge with bacteria capable of accumulating orthophosphate in excess of the normal metabolic requirements of the cell. Many bacteria isolated from activated sludge, i.e. *Acinetobacter* sp., *Aeromonas hydrophila*, *Pseudomonas* sp., *Moraxella* sp., *Enterobacter* sp., *Xanthobacter* sp., *Comamonas–Pseudomonas* group, *Zoogloea ramigera*, some filamentous bacteria, coryneform bacteria and some

Gram-positive bacteria, have been reported to accumulate polyphosphates (Dienema *et al.* 1980; Lötter 1985; Lötter and Murphy 1985; Suresh *et al.* 1985; Streichan *et al.* 1990).

1.2.7 Microbial ecology of activated sludge

The ecological study of the activated sludge process is of basic importance for determining which microorganisms have a significant role in this process.

In activated sludge, the relative number of each species is determined by availability of oxygen, pH and mode of mixing, which determines the growth rate of the various species. Thereby the dominant bacteria will be those capable of most effectively utilizing the organic wastes and having the ability to flocculate to ensure their retention in the system. The nature of the bacteria is hence determined by the nature of organic wastes and by conditions in the process such as anaerobic, anoxic and aerobic conditions (Buchan 1984).

The microbial community of activated sludge consists of bacteria, protozoa, fungi, algae and filamentous organisms. Fungi and algae are not considered so important, whereas protozoa, filamentous organisms and bacteria actively participate in the biological treatment of wastewater by the activated sludge system. The protozoa, responsible for 'grooming' the zoogloeal mass by grazing on it, are consumed by other organisms present in the system. In this way, a food chain is created and at each stage of the food chain a fraction of the original material is removed from the system as carbon dioxide (Boyd 1984). The ciliates such as *Vorticella, Opercularias* and *Epistylis* play an important role in activated sludge systems (Shapiro 1967). Electron microscopic investigations showed that certain amoebae contained large numbers of the phosphate-accumulating bacterial cells. It seemed as though these amoeba preyed only on bacterial cells (Buchan 1983). The filamentous organisms, at a lower frequency of occurrence, can contribute to the proliferation of large, firm, activated sludge flocs exhibiting good settling properties. When the occurrence of filamentous populations exceed a certain level (e.g. 10^7 µm/ml or 10^4 m/g, according to Jenkins *et al.* (1986), the following settling and compaction problems can be observed:

- low zone settling velocities;
- excessive volume of the settled activated sludge; this is diluted and at once affects the operation of the aeration basin and the processing of waste activated sludge (Wanner 1993).

The major biological problems directly linked with filamentous populations are bulking and foaming (Jenkins *et al.* 1986). Both cause solid separation problems in that the final liquid effluent is contaminated with bacteria-rich sludge (Blackall 1993). Bulking of activated sludge is caused by the proliferation of filamentous microorganisms, whereas their absence causes foaming (Eikelboom 1993) and a pinpoint floc (Jenkins *et al.* 1986). The filamentous microorganisms most frequently found at present are: *Microthrix parvicella, Halicsomenobacter,* type 021 *hydrossis*, type 0092, type 1701, *Nocardia amarae, Nocardia pinensis, Sphaerotillus natans, Nostocoida limicolla*, type 0961, 021N/*Thiothrix*, 0041/0675 and *Rhodococcus* spp. (Jenkins *et al.* 1986; Blackall 1993; Duchene 1993; Eikelboom 1993; Wanner 1993). Excessive growth of *Nocardia erythropolis* and *Microthrix parvicella* lead to scum formation, which can produce offensive odours; if discharged in the final effluent, a significant increase in COD and suspended solids content can result.

The wastewater liquor in activated sludge systems consists of a heterotrophic bacterial community. However, the lack of suitable media that would support the growth of all viable nutritional types of bacteria in activated sludge (Banks and Walker 1976) has been a limiting factor in the identification of the bacterial community in activated sludge systems. In spite of this handicap, several bacterial studies have been done in the system. *Zoogloea ramigera* has been accepted as the predominant bacterium in activated sludge responsible for the stabilization of the organic matter as well as for the production of floc (McKinney and Weichlein 1953). Not only do bacteria belonging to the genus *Zoogloea* predominate in activated sludge, but also the Gram-negative bacteria belonging to the genus *Comamonas* (Dias and Bhat 1964). Recently, the bacterial population in activated sludge nutrient removal systems has been studied by many researchers, and the organisms investigated in this system include species of *Acinetobacter, Moraxella, Flavobacterium, Pseudomonas, Vibrio, Achromobacter, Alcaligenes, Enterobacter, Serratia, Proteus, Aeromonas, Proteobacter, Xanthobacter, Aerobacter, Klebsiella, Bordetella, Citrobacter, Shigella, Pasteurella, Yersinia* spp, *Escherichia intermedium* and *Bacillus cereus* (Harold 1966; Shapiro 1967; Fuhs and Chen 1975; Deinema *et al.* 1980; Shoda *et al.* 1980; Brodisch and Joyner 1983; Gersberg and Allen 1984; Lötter 1985; Lötter and Murphy 1985; Suresh *et al.* 1985; Venter *et al.* 1989; Streichan *et al.* 1990; Auling *et al.* 1991; Kavanaugh and Randall 1993).

Acinetobacter were frequently isolated from activated sludge plants operated to achieve the biological removal of phosphate (Fuhs and Chen 1975; Nicholls and Osborn 1979; Buchan 1980; Deinema *et al.* 1980; Brodisch 1985; Cloete *et al.* 1985; Lötter 1985; Suresh *et al.* 1985). A number of reports have suggested that the dominant organism is either *A. calcoaceticus* or *A. lwoffi* (Fuhs and Chen 1975; Buchan 1983; Marais *et al.* 1983; Lötter 1985; Lötter and Murphy 1985; Murphy and Lötter 1986). This organism therefore became the model organism for studying mechanisms of biological removal of phosphate. The observations of Deinema *et al.* (1980) indicated that bacteria from a biological phosphate removal plant that grew in the presence of acetate agar and accumulated polyphosphate all belonged to the genus *Acinetobacter.* By using the viable plate count and the Analytical Profile Index (API)-20E techniques for biochemical characterization and identification, Buchan (1980) found that *Acinetobacter* was the dominant organism in activated sludge. Lötter (1985), studying a full-scale plant and using a fluorescein isothiocyanate-labelled fluorescent antibody against *Acinetobacter* spp., followed by the API-20E identification system, indicated that 56% of the bacterial colonies that grew on a basic medium with acetate as carbon source were identified as *Acinetobacter*. Fluorescent antibody (FA) techniques were also employed to estimate the number of *A. calcoaceticus* and *A. phosphodevorans* in activated sludge (Cloete *et al.* 1985). In all cases, the number of *Acinetobacter* exceeded 10^6 cells/ml and were highest in the aerobic zones. The total number of bacteria in activated sludge counted with the acridine orange (AO) staining technique, compared with the FA number of *Acinetobacter*, indicated that *Acinetobacter* constituted less than 10% of the total bacterial population in activated sludge (Cloete and Steyn 1987). Brodish and Joyner (1983) used the API system to determine the composition of the bacterial community in the anaerobic, anoxic and aerated stages of three biological phosphate removal plants and two laboratory scale units. In contrast with the findings of Buchan (1980) and Lötter (1985), results revealed that *Acinetobacter* were present in minor proportions, whereas *Aeromonas* and *Pseudomonas* were more active and constituted more than 50% of the total aerobic microbial population. Direct staining for polyphosphate granules, cultured in ^{32}P-labelled inorganic phosphate before autoradiography, and tests for resistance to growth with the ATPase inhibitor dicyclohexyl carbodiimide (DCCD), were used to isolate organisms involved in the uptake and subsequent release of inorganic phosphate from wastewater sludge (Suresh *et al.* 1985). Results have shown that *Pseudomonas vesicularis* was very active and could store more phosphate than *Acinetobacter* (Suresh *et al.* 1985).

Evidence from the above experimental studies proved that *Acinetobacter* spp. were not the dominant, nor the only, organism responsible for enhanced biological removal of phosphate. Other bacteria such as *Aeromonas* (Brodisch and Joyner 1983), *Klebsiella pneumoniae* (Gersberg and Allen 1984), *Pseudomonas* and *Escherichia coli* were also implicated in the removal of phosphate (Cloete *et al.* 1992).

1.2.8 Applications of the activated sludge process

There are many successful applications of biological nutrient removal in activated sludge systems: Phoredox process, A/O process, UCT process, Biodenipho process, BB process, modified oxidation ditch process, Phostrip process and Bardenpho process, to name a few. These will be discussed briefly.

1.2.8.1 The A/O process

The A/O (aerobic/oxic) process was developed in the U.S.A. This process is one of the simplest biological phosphorus removal systems. The returned activated sludge is mixed with the incoming wastewater and this mixed liquor passes through an anaerobic zone and then through an aerobic zone. Phosphorus accumulates in the sludge and is removed by the surplus sludge. The A/O process is used for high-loaded activated sludge systems. For low-loaded activated sludge systems an anoxic zone is introduced between the anaerobic and the aerobic zones for denitrification (Van Strakenburg *et al.* 1993).

1.2.8.2 The Phoredox process

The Phoredox (phosphorus reduction oxidation) process was designed by Barnard (1976). It is a modification of the Bardenpho process to facilitate phosphorus and nitrogen removal. It consists of an anaerobic zone as well as anoxic and aerobic zones. The key process in this system is the anaerobic fermentation zone with a short retention time (1–3 h). The anaerobic zone encourages the production of low molecular mass fatty acids to promote the growth of *Acinetobacter*. Phosphorus uptake occurs in the first aerobic zone. The function of the second anoxic zone is to denitrify the incoming nitrate from the first aerobic zone by using the endogenous carbon source. The final aeration zone is intended for sedimentation to stimulate the release of nitrogen and to improve sludge settleability (Streichan *et al.* 1990; Van Strakenburg *et al.* 1993).

1.2.8.3 The UCT process

The UCT (University of Cape Town) process was developed by Ekama *et al.* (1984). This process is derived from the Phoredox process. The modification includes three internal recirculations resulting in a decrease in the nitrate concentration. The returned activated sludge is recycled into the anoxic zone instead of the anaerobic zone. The other two recirculations are located between the aerobic and the anoxic zones, and between the anoxic and the anaerobic zones (Van Strakenburg *et al.* 1993).

1.2.8.4 The Biodenipho process
The Biodenipho (biological denitrification phosphorus removal). This process is characterized by an alternating supply of two coupled aeration tanks with anoxic and aerobic phases. To promote the growth of phosphate-accumulating bacteria, this is preceded by an anaerobic zone. The anaerobic zone receives the incoming wastewater and return sludge. It normally consists of three compartments to promote the plug-flow character. By alternating wastewater supply, optimum nitrification and denitrification occurs. Recirculation of nitrified water to the anoxic zone is excluded because both processes take place in the same tank (Van Strakenburg *et al.* 1993).

1.2.8.5 The BB process
The BB process is characterized by the activated sludge undergoing anaerobic, anoxic and aerobic phases over time as a result of alternating aeration. During the anaerobic/anoxic phase the mixed liquor is not mixed but settles to the bottom of the first aeration tank. The supernatant enriched with nitrate does not inhibit phosphate release in the first aeration tank. The nitrification and denitrification take place in the first and second aeration tanks. During periods of aeration in the first and second aeration tanks, phosphate-accumulating bacteria take up phosphate (Van Strakenburg *et al.* 1993).

1.2.8.6 The modified oxidation ditch process
In a carousel or oxidation ditch process the activated sludge mixed liquor flows continuously around a loo-type channel. An aeration system ensures the aeration of the activated sludge. With low oxygenation it is possible to create an aerobic zone capable of nitrification, immediately downstream of the reactor and an anoxic zone some distance upstream of the reactor. The discharging of the influent at the upstream limit of the anoxic zone favours the use of some of the wastewater carbon source for denitrification. An anaerobic zone positioned in front of the ditch, where the return sludge meets the influent, ensures an optimum combination of phosphorus and nitrogen. For biological phosphate removal purposes, the following modifications must be considered: the first two sections of the circuit serve as the anaerobic zone, the recirculation flow must go from the last to the third section, and the section of the circuit for sedimentation serve as anoxic zones (Van Strakenburg *et al.* 1993).

1.2.8.7 The Phostrip process
The Phostrip process is a combined biological and chemical process for the removal of phosphorus from wastewater. The Phostrip process consists of an activated sludge reactor into which recycled sludge, stripped of phosphorus is introduced with an influent feed. This side stream process relies on the ability of sludge microorganisms to release phosphorus in a more concentrated form in a small anaerobic phosphate stripped tank. Aerated mixed liquor enters a settler, where activated solids are separated and the supernatant is discharged. The solids are passed through an anaerobic clarifier where phosphorus is released from the organisms into the supernatant. The P-stripped sludge is returned to the head of the activated sludge reactor, while the P-enriched secondary supernatant is treated separately with lime to precipitate phosphorus (Streichan *et al.* 1990).

1.3 Biofilm reactors

Microbial film process technology has developed rapidly in the past two decades and is often used for wastewater treatment. A biofilm refers to a complex structure of cells and cellular products, such as extracellular polymers, attached to a solid surface or substratum. Immobilization of bacteria or other microorganisms as biofilms can be achieved either naturally or artificially. Under specific natural conditions, microorganisms form biofilms spontaneously. Bacterial growth on surfaces is a natural phenomenon because most microorganisms in nature are associated with solid surfaces. In streams and rivers a large proportion of the microbial activity occurs in attached films. This microbial activity is responsible for the self-purification capacity of the rivers by transformation and degradation of natural and man-made compounds in the water (Tijhuis 1994). The artificial immobilization methods of microbial cells for microbial film process purposes can be classified broadly into three categories as carrier binding, entrapment and cross-linking methods.

1.3.1 Configuration of biofilm reactors

Practical biofilms exhibit considerable periodic fluctuations in biofilm thickness and biomass per support area, because increase by growth and decrease by erosion and by sloughing are taking place continuously. Erosion can be defined as the continuous loss of small portions from a biofilm body, and sloughing refers to sporadic detachment of large fragments from a biofilm body. In natural biofilms, besides diffusion and growth processes, for instance, biofilm detachment occurs, and the biofilm surface is not rigid. Typical biofilms develop in a sigmoidal manner, which is divided into five phases. These are: microorganism adhesion to a surface, exponential growth phase, declined growth phase, plateau, and repeats of sloughing and renewal. The nature of biofilms is that the

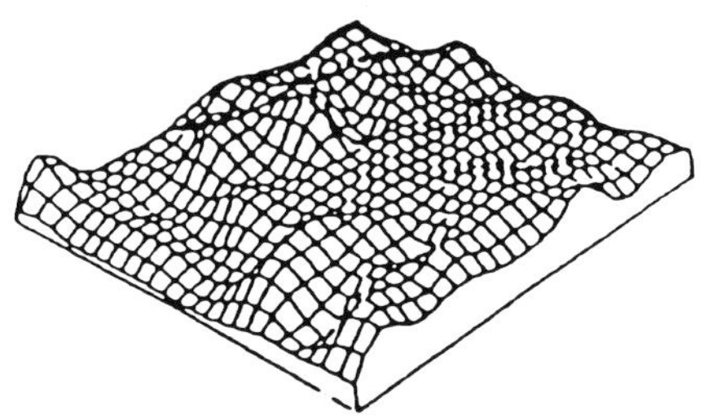

Figure 1.7. Surface configuration of a biofilm after 48 hours.

organisms live very closely attached to each other, therefore providing resistance to mass transfer of products out of the biofilm. Biofilms are not always thin films that cover flat solid walls. Biofilms very often become filament-shaped in conditions of high velocity flow (Nagaoka and Sugio 1994). The structure of biofilms is influenced by external shear, internal diffusion profiles and microbial competition. Shear forces on the biofilm determine the film's thickness and also the structure of the biofilm surface. In most biofilm reactor systems shear forces are not uniform. This results in a non-uniform biofilm formation in such reactors. Figure 1.7 describes the surface configuration of a typical biofilm 48 h after incubation.

1.3.2 Principles of biofilm reactors

Biofilm grown on the surface of a medium ingests substrate as organic matter, oxygen, trace elements, etc., required for biological activity, from the liquid phase with which it is in contact. According to Zhang and Bishop (1994), the overall process of substrate utilization by biofilms can be briefly described by three main steps. These are: substrate transport from the bulk liquid to the external biofilm surface, diffusion of substrate within the biofilm, and substrate consumption by microorganisms within the biofilm. The first step is of most importance although it is a purely physical phenomenon because, under the steady-state condition, the rate of the overall reaction resulting from steps 2 and 3 is always equal to the rate of the first step.

Colloidal or suspended organic matter cannot diffuse into the biofilm directly. These must be hydrolysed to low molecular mass organic substances before the diffusion process can take place. The end products of metabolism are in reverse transferred into the liquid phase. When any nutrient essential to microorganisms is not supplied, biological reactions will be affected. Therefore, if any one of these nutrients is exhausted at a certain depth of biofilm, biological reactions will not occur in the deeper portion. Thus the substance exhausted first determines the effective depth of a biofilm (Iwai and Kitao 1994). The above substance is termed the *limiting factor*. Nevertheless nutrients such as nitrogen, phosphorus and other trace metals are not likely to become limiting factors because they are usually in excess in wastewaters.

1.3.2.1 Phosphate removal by biofilms

Most investigations for phosphorus removal on biofilters were based on physico-chemical precipitation. Biological removal of phosphate on biofilm reactors remains limited to a few laboratory-scale experiments. However, because of characteristics such as compactness, operational flexibility and ease of operation, the biofiltration process presents several advantages for biological removal of phosphate. Experiments with floating upflow biofilters showed that it is possible to adapt a biofilter process to obtain biological removal of phosphate by introducing one or more anaerobic contact periods during the filtration period between two backwash procedures of the same biofilter. Moreover, Goncalves *et al.* (1994) have reported simultaneous nitrification, denitrification and biological phosphate removal of the settled wastewater on a pilot plant scale.

1.3.2.2 Nitrification and denitrification

Watanabe *et al.* (1994) have reported oxidation of organic matters and suspended solid nitrification and denitrification by an upflow aerated biofilter (UAB). They concluded that UABs have a great potential ability for oxidation of organic matter, stabilization of suspended solids, and nitrification due to a unique aeration mechanism giving high dissolved oxygen concentrations with relatively low aeration rates.

1.3.3 Microbiology of biofilm reactors

Biofilms are known as complex structures of microbial cells and extracellular polymers. All microorganisms natural to wastewaters (both municipal and industrial) can also be found in biofilms fed with wastewater. However, the diffusion gradients inside the biofilm will influence the growth rate and competition of different microorganisms. It is generally believed that slow-growing organisms (e.g. nitrifiers or microorganisms growing on xenobiotics) will be found deeper inside a biofilm than fast-growing heterotrophs. The distribution of microorganisms in biofilms can be studied by direct observations with confocal laser microscopy. Van Loosdrecht *et al.* (1994) have studied the population distribution of nitrifying and heterotrophic bacteria in biofilm particles from biofilm airlift reactors and reported the existence of a layered structure of nitrifying and heterotrophic bacteria in these biofilms. Their findings indicated a non-uniform distribution of species inside the biofilm.

1.3.4 Applications of biofilm reactors

Several reactor types have been developed to exploit the benefits of biofilm processes. According to Tijhuis (1994), biofilm reactors can be subdivided roughly into two categories: biofilms on a fixed substratum and biofilms on suspended particles.

1.3.4.1 Reactors with biofilms on fixed substrata

Reactor systems with fixed substrata are widely used for the treatment of different kinds of wastewater. The shear forces acting on the biofilms are low, so the biofilms grow out very thickly, up to 1 cm for instance in trickling filters and rotating biological contactors, but only a small part of the biomass is aerobically active.

Trickling filters. Trickling filters are the oldest biofilm reactors for wastewater treatment. A trickling filter consists of a bed of support material over which the wastewater is uniformly distributed at surface irrigation rates of 0.5–2.5 m^3/m^2 per hour (Figure 1.8). The wastewater percolates over the biofilm growing on the carrier material. In the past, most trickling filters used rock media, with a diameter of 25–100 mm. This resulted in a biofilm specific surface area of 50–100 m^2/m^3. The height of such a filter is in general *ca.* 2 m. Nowadays, low-density plastic media such as polystyrene or polyvinyl chloride are used, which both have a higher specific surface area of *ca.* 300 m^2/m^3. The reactor can then attain a height of up to 12 m.

Rotating biological contactors. A biological contactor consists of a series of closely spaced circular plastic disks of polystyrene or poly(vinyl chloride) (Figure 1.9). The disks are partly submerged in the wastewater. By gentle rotation of the disks the biofilms are alternately exposed to the contaminants in the wastewater and oxygen in the air. Biofilm attaches to the surfaces of the disks during operation. The disk rotation affects oxygen mass transfer and maintains the biomass in the aerobic condition. The size of the rotating biological contactor depends on the disk diameter, which cannot be too large owing to mechanical limitations.

Submerged biofilters. A typical submerged biofilter consists of a reactor that is packed with a filter medium to which microorganisms can become attached (Figure 1.10). The filter material is continuously submerged in the wastewater. Air is introduced beneath the medium to aerate the wastewater. A large variety of packing materials are used in submerged biofilters such as expanded clay and polystyrene. The diameter of these materials is *ca.* 5 mm. Two modes of operation of a biofilter are possible: downflow and upflow. The filters must be regularly backwashed to remove produced biomass

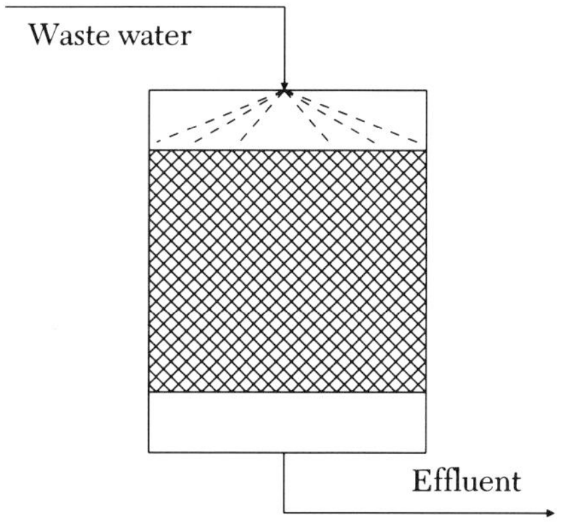

Figure 1.8. Trickling filter.

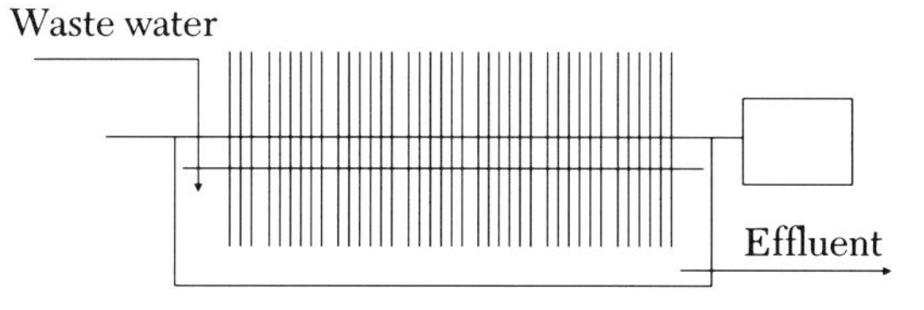

Figure 1.9. Rotating biological contactor.

from the filter material and to prevent pressure build-up and clogging of the filter medium.

1.3.4.2 Reactors with biofilms on suspended particles

Mechanical stirring cannot be applied for the suspension of the biofilm particles because of the detrimental effect of the stirrer on the biofilms. Fluidized-bed reactors or airlift reactors are used to suspend the particles. These reactors present a high biofilm surface area, up to 3,000 m^2/m^3 with active biomass concentrations.

Fluidized-bed reactor. By recirculation of the effluent, the particles in this reactor are fluidized and, for aerobic processes, the reactor contents are aerated by pure oxygen or, possibly air (Figure 1.11). For the treatment of high-strength wastewaters the recycle rate must be high to bring enough oxygen into the reactor. When there is an oxygen demand of the incoming water of 500 mg/l, the recycle flow rate with the use of air or pure oxygen is 62 and 12.4 times the influent flow rate, respectively. To overcome these high recirculation rates the reactor might be aerated directly, which is called a three-phase fluidized-bed reactor. In the fluidized-bed reactor the biofilm particles are segregated; larger particles tend to be found at the top of the reactor. Owing to the low detachment forces, biofilms grow in these reactors, requiring additional measures for controlling biofilm thickness.

Airlift reactors. Airlift reactors consist of a liquid pool divided into two distinct zones, of which one is usually sparged with gas (Figure 1.12). The different gas holdups in the gassed and ungassed zones result in different bulk densities of the fluid in these regions; this

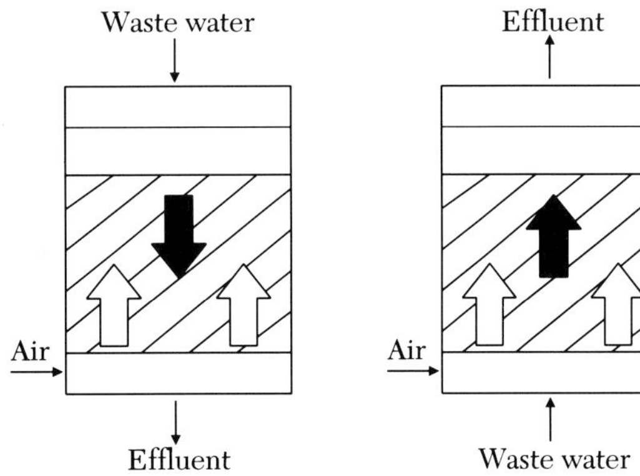

Figure 1.10. Submerged biofilters.

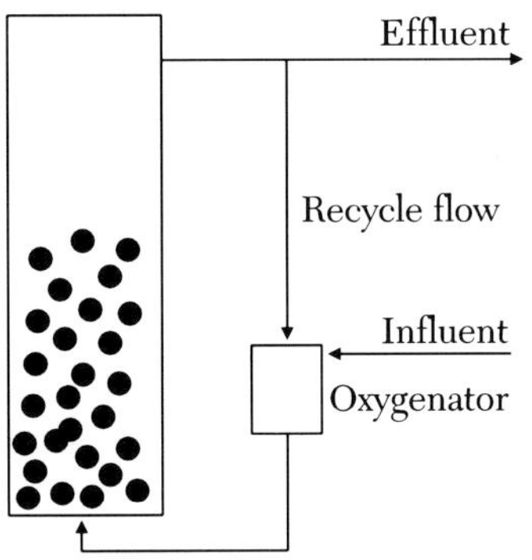

Figure 1.11. Fluidized-bed reactor.

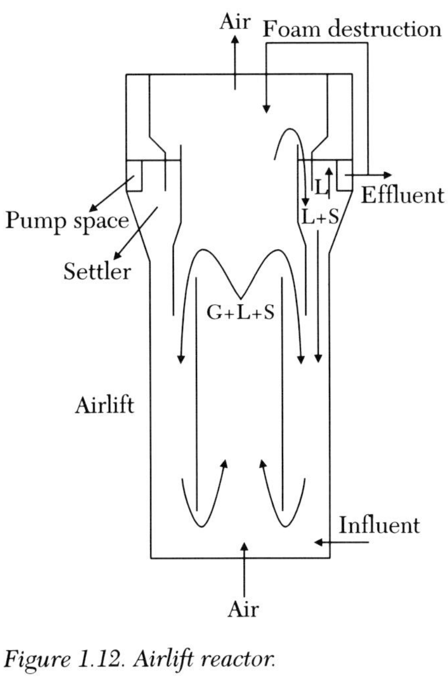

Figure 1.12. Airlift reactor.

causes circulation of the fluid in the reactor by a gas–liquid upflow termed the riser; the region containing the downflowing fluid is termed the downcomer. Practical applications of airlift reactors depend on the ability to achieve the required rates of momentum, heat and mass transfer at acceptable capital and operation costs.

A typical application of an airlift reactor for wastewater treatment is 'the deep shaft'. The advantages of such a treatment method over conventional methods include the fact that it takes considerably less space, has minimal visual impact and greatly reduces smells (Redman 1987). The basic point of the deep shaft is that it is an aeration tank built vertically into the ground, rather than as a pool at ground level. The sewage goes down one side of the shaft and comes up the other. The sewage is aerated and kept in motion by the injection of air on the downcomer and riser.

1.4 Anaerobic digestion processes

Anaerobic digestion is one of the oldest means of wastewater treatment. By far the primary use of anaerobic digestion is in the stabilization of suspended organic matter (Grady and Lim 1980). However, the rapidly rising cost of fossil fuels has generated increasing interest in producing methane from organic matter through anaerobic digestion. Anaerobic digestion refers to a biological engineering process by which a complex feedstock is converted into a range of simpler compounds including methane by microorganisms in the absence of oxygen (Hughes 1979). Because the term anaerobic implies the absence of oxygen, usually by the strict exclusion of air, all reactions in the anaerobic digestion are performed by anaerobic microorganisms, mainly bacteria. The physiology and biochemistry of these anaerobic microorganisms therefore determine the special characteristics of the engineering systems commonly known as digesters. From an energy consumption viewpoint, anaerobic digestion seems to be by far the most economic biological wastewater treatment process. This is because the anaerobic treatment of wastewaters can result in a substantial decrease in power requirements. Moreover, when substantial quantities of biodegradable organic wastes are involved, the resulting quantity of methane can produce energy in excess of that required for the wastewater treatment systems (Pfeffer 1979). Hashimoto and Chen (1979), studying the overall economics of anaerobic digestion, concluded that this process is technically feasible, can produce enough energy to satisfy the energy requirements of the digestion system and is economically attractive in certain instances.

1.4.1 Configuration of anaerobic digestion processes

There are three basic configurations for the methane fermentation process through anaerobic digestion.

The most common system is a single mixed reactor without solid returns, which has been applied to concentrated slurries with a significant quantity of non-biodegradable solids, both organic and inorganic. This system has few or no applications to wastewater treatment because of the volume required to achieve the required retention times of 15–30 days. It is used essentially for the treatment of organic sludges produces by other wastewater treatment processes.

The second process configuration is commonly known as the anaerobic contact process. The flow pattern is essentially the same as that used in the activated sludge process. A mean cell residence time greater than the hydraulic retention time is achieved by settling the solids from the effluent and recirculating concentrated sludge back to the fermenter. Because waste stabilization efficiency depends primarily on the mean cell residence time, it is possible to obtain a high treatment efficiency without having excessive digester volumes. The separation of the solids from the effluent stream greatly improves the process's BOD and efficiency of removal of suspended solids (Pfeffer 1974). To correct the poor separation of solids in the clarifier, the flow from the fermenter to the clarifier is passed through a degasifier. This strips the dissolved gases from the slurry, resulting in improved settling. The anaerobic contact process has been applied to some industrial wastes. However, it has not been widely used, even though it is more economical than the activated sludge system.

The third process, developed by Young and McCarty (1969), consists of a tower filled with packing material. The packing provides a surface to which the microorganisms attach. The need for solids separation and sludge recycling is therefore eliminated. The community of microorganisms that develop in the tower is in equilibrium with the organic loading and the hydraulic retention time. Because this is an upflow column, suspended matter tends to accumulate in the column. The microbial metabolic activity will tend to establish an equilibrium between the rate of biodegradable material depositions and the rate of bioconversion. Only the non-biodegradable solids will accumulate and this can become a problem if the content of these solids in the wastewater is high.

1.4.2 Principles of the anaerobic digestion process

The anaerobic digestion of organic compounds to methane and carbon dioxide is a multistep process involving different physiological groups of microorganisms (Pretorius 1994). However, according to Pfeffer (1979) the process can be considered in its simplest form to be a three-stage process.

In the first stage, a group of anaerobic microorganisms, primarily cellulolytic bacteria, act on the organic polymers. The reaction is an enzymic hydrolysis of the polymers to the individual monomers.

In the second stage, these monomers are fermented to various intermediates, primarily acetate, propionate and butyrate. Additional acetate is produced by a second group of microorganisms, commonly called acetogenic bacteria.

In the third stage, which is probably the most important, acetic acid becomes the substrate for a group of strictly anaerobic methanogenic bacteria. These bacteria ferment acetic acid to methane and carbon dioxide. This methane, along with the methane formed by bacteria that reduce carbon dioxide by using hydrogen gas or formate produced by other species, accounts for the methane produced in this process.

It must be pointed out that methanogenic bacteria seem to be extremely sensitive to certain environmental factors, in particular oxygen, pH, ammonia, temperature and nutrients.

1.4.2.1 Factors affecting the anaerobic digestion process

Effect of pH. Although several acid systems are involved in anaerobic digesters, the carbonic system has been reported to be by far the most important governing the value and stability of pH (Capri and Marais 1975; Li and Sutton 1983; Pretorius 1994). A desired pH can be obtained either by adding alkalinity to the influent or by abstracting acidity (CO_2) from the digester (van Haandel 1994). The optimum pH range seems to be between 6.5 and 7.5. When the pH falls below this optimum, significant inhibition of the methanogenic bacteria occurs. At a pH of *ca.* 6.2, the acid conditions exhibit acute toxicity to these bacteria.

Effect of ammonia. Ammonia, particularly in the form of NH_3, is inhibitory when present in high enough concentrations. At concentrations between 1,500 and 3,000 mg per litre of total ammonia nitrogen and a pH greater than 7.4, NH_3 can become inhibitory. However, at a concentration above 3,000 mg/l, the ammonium ion itself becomes toxic regardless of pH (McCarty and McKinney 1961).

Effect of oxygen. As obligate anaerobes, methanogenic bacteria are inhibited by a small concentration of oxygen. Maintenance of a low oxidation–reduction potential is thus essential for the successful operation of anaerobic digestion processes.

Effect of temperature. Temperature is a very important operational factor in anaerobic digestion processes. As temperature increases, the rates of reaction proceed much faster and this results in more efficient operation and lower retention time requirements. Because the methane requirements to maintain thermophilic temperatures in most digesters are excessive and uneconomic, sewage sludge digestion systems are operated in the mesophilic temperature range (35–40 °C).

Nutrient requirements. The bacteria responsible for the waste conversion and stabilization in the anaerobic process requires nitrogen, phosphorus and other nutrients in trace quantities for optimal growth. Municipal wastewater usually contains all the required nutrients in adequate quantities. Other elements such as sodium, potassium, calcium, magnesium and iron have only stimulatory effects at low concentrations. However, all the above elements could exhibit inhibitory effects at too high concentrations.

1.4.3 Microbiology of anaerobic digestion processes

Our present understanding of bacterial populations in anaerobic digesters is rather limited and is based largely on the analysis of bacteria isolated from sewage sludge and digesters of animal manure, or from the rumen of cud-chewing animals. The digester microbial population is composed of several major trophic groups, namely: hydrolytic bacteria that catabolize saccharides, proteins, lipids, and other minor chemical constituents of biomass; hydrogen-producing acetogenic bacteria that catabolize certain fatty acids and neutral end products; homoacetogenic bacteria that catabolize one-carbon compounds or hydrolyse multicarbon compounds to acetic acid; and methanogenic bacteria that catabolize acetate and one-carbon compounds to methane. The coordinated activity of these trophic groups as a whole ensures process stability during anaerobic digestion.

The generic identification of the anaerobic bacterial populations in sewage sludge digesters have indicated the presence of: Gram-negative rods, *Eubacterium*, *Clostridium* (as hydrolytic bacteria); Gram-negative rods (as hydrogen-producing acetogenic bacteria); *Clostridium*, *Acetobacterium* (as homoacetogenic bacteria); *Methanobacterium*, *Methanospirillum*, *Methanococcus*, *Methanosarcina* (as methanogens); and *Desulfovibrio* and *Desulfotomaculum* (as sulphate reducers). Pretorius (1994) has indicated that the prevailing COD (short-chain fatty acids) concentration has a marked effect on the microbial population selected. When the average COD was 2,541 mg/l during the start-up phase, *Methanococcus* type bacteria dominated, but when the COD was reduced to 783 mg/l, *Methanothrix* bacteria dominated.

1.4.4 Applications of anaerobic digestion

A worldwide review of anaerobic digesters indicated that these could be divided into three groups, namely rural-type digesters, classical industrial digesters and newly conceived digesters.

1.4.4.1 The rural-type digesters

These are the simplest and most robust digesters, so termed because they are intended for rural areas in developing countries as well as in industrialized countries. The accent is laid on the ease with which they can be operated, and on their reliability, rather than on maximum methane outputs. In this section we discuss the French-type, the Chinese-type and miscellaneous rural digesters.

The French-type digester. It consists of a batch digester intended to metabolize semi-solid material. The first one was constructed in 1941 at the Agricultural Institute of Algeria. It has a capacity of 10–50 m^3. The gas tightness of the cover is ensured by a water-ring. The cover can glide upwards and downwards. This type of digester also serves as a gasometer.

Manure is mixed with straw in proportions up to 1:9 to ensure the appropriate C/N ratio of *ca.* 30 necessary for proper methanogenesis. In a first step, the mixture of manure and straw is subjected for 4–7 days to an aerobic composting-like fermentation. The purpose of this aerobic pretreatment seems to be the oxidation of easily fermenting compounds; otherwise the latter compounds might cause an excessive anaerobic acidogenic fermentation to take place in the digester, which would lower the pH of the mixed liquor and prevent proper methanogenesis. In the second step, the pretreated mixture is seeded with part of the preceding anaerobic digestion and subjected to an anaerobic fermentation. The temperature is maintained between 30 and 35 °C, and the pH between 7.5 and 8.5.

The Chinese-type digester. It consists of a batch digester intended to digest semi-liquid material. Its building was initiated around 1958 but has mainly been developed during recent years. It has been reported that about seven million such digesters of *ca.* 6–8 m^3 capacity are currently in operation. The methanogenic substrate is animal and human manure and urine. These digesters are operated on a basis intermediate between batch and semi-continuous. They are fed once a day and this ensures mixing. A similar amount of decanted mixed liquor is also removed once a day. The suspended solids accumulate in the bottom part of the digester. Therefore the digesters have to be emptied two or three times a year. One fifth of the preceding anaerobic sludge serves as seed for the following digestion.

Miscellaneous rural digesters. A number of full-scale digesters intended for farm wastes have been developed. Because these resemble the classical completely mixed digesters, they will not be dealt with in this section. Our focus will rather be on the displacement-type digester.

The concept of the displacement-type digester started in South Africa and ended in California. It consists of a horizontal cylinder loaded at one end and discharged at the other.

The digester is no longer of the completely mixed type, but rather of the piston plug-flow type. The solids retention times are between 35 and 40 days. Fresh manure has to be diluted with urine or water for proper handling. The volumetric load is *ca.* 2 kg volatile solids/m^3 of digester per day.

In the Philippines, for instance, the Agro-industrial Division of Liberty Flour Mills operates displacement-type digesters with a total capacity of 1,800 m^3. The biogas production rate obtained is 0.7 m^3 gas/m^3 of digester per day.

1.4.4.2 The classical industrial digesters

These are of the completely mixed type and are operated in a continuous or at least semi-continuous mode. The classical industrial digesters are usually egg-shaped or cylindrical in form. These are built from concrete or metal. Both methanogenic substrate and digestion mixture used are liquid suspensions. The digesters are operated on a continuous basis. An essential characteristic is that they belong to the completely mixed type. Mixing occurs by mechanical stirring, either by the recirculation of the liquid suspension or by stirring by the gas mixture evolved.

A very large number of pilot digesters (1–30 m^3) are in operation all over the world, although only a limited number of full-scale digesters are reported to be in operation. In the U.S.A., for instance, many sewage anaerobic digesters are in use. As far as other types of wastes are concerned, projects involving the conception, construction and demonstrative operation of a full-scale completely mixed digester are in progress. Limited information is available on eventual full-scale digesters in operation in Europe. However, anaerobic digesters using excess sludge from wastewater treatment are very common. These are primarily intended for sludge stabilization rather than methane production.

1.4.4.3 Newly conceived digesters

These are designed on the basis of recent findings in the fundamental biochemistry and microbiology of methanogenesis. The overall methanogenesis of complex substrate is a multi-step process. The idea of splitting the anaerobic digestion process into two steps, each of which can be optimized separately in an overall system that is more economical than the single-step process, has been put forward.

A separate first step, aimed at improving the accessibility or liquefaction of the complex organic substrate, is required. This can be either of a physico-chemical nature, e.g. a heat treatment, or of an enzymic nature, e.g. a cellulolytic or hemicellulolytic treatment. However, it is as yet not definitely clear whether it is advantageous to split the process into two steps: a first one optimized for acidogenesis and a second optimized for methanogenesis. Whenever the two-step process is chosen, acidogenesis can be performed in anoxic conditions and a simple open digester used, which at the limit can be a pond or a ditch. Neither the maximum growth rates of methanogenic bacteria nor the specific production rates of methane by methanogenic bacteria seem to be limiting factors for the process. Mixing is required only to mix the load with the mixed liquor and to break the top scum layer. Nyns *et al.* (1979), reporting on a worldwide review of digesters, concluded that a digester intended for industrial purposes must lay the accent on the efficiency of bioconversion, on both thermodynamic and kinetic grounds. A two-stage process consisting of a first step optimized for accessibility–liquefaction and a second step consisting of an upflow piston plug-flow methanogenic digester seems to be the most promising combination.

1.5 Lagoon systems

Lagoon systems provide a low-cost sewage treatment system for small communities. About five types of lagoon system are used for wastewater treatment, namely the aerobic, facultative, anaerobic, aerated and maturation lagoons. The applicability of a lagoon system depends on the strength of the waste, the land area available and the quality of the effluent required. A lagoon refers basically to a hole in the ground where the wastewater is stored to undergo natural purification, which is brought about by the bacterial activity with symbiotic action from algae and other organisms (Ouano 1983). The biological diversity in a lagoon system is higher than in other biological wastewater treatment processes. As a result, the lagoon system is a very stable process. Owing to the complexity of the biological reaction and interactions that take place in this process, the rationalization of design criteria has become a problem. However, the system has been accepted in many areas as a satisfactory method for disposing of household sewage and some types of industrial waste. The following section gives a scanning view of the different lagoon types.

- *Aerobic lagoon.* An aerobic lagoon might be intended for maximum oxygen or algal production. An aerobic condition prevails throughout the pond depth, which is very shallow at 0.3–0.4 m. This system can remove about 80–95% of the wastewater-soluble BOD. The aerobic lagoon system is used primarily for algal protein production because it requires a very large extent of land per unit BOD stabilized compared with the facultative or anaerobic system.
- *Facultative lagoon.* A facultative lagoon is slightly deeper than the aerobic lagoon. Its

top layer is always aerobic because the bottom is anaerobic. Between those two layers, the dissolved oxygen concentration could vary from supersaturation in mid-afternoon to negligible content during the early hours of dawn. The facultative lagoon system is more widely used than the aerobic and anaerobic lagoon. Where land is a limitation, facultative lagoons can be used to treat the effluent from anaerobic lagoons to achieve the effluent quality standards.

- *Anaerobic lagoons.* Anaerobic lagoons are very much deeper than facultative lagoons. Owing to their great depth, the amount of oxygen supplied from the atmosphere is insufficient to maintain aerobic conditions except for a very thin layer at the top of the water during start-up. Later, a scum layer is formed above the wastewater that prevents algae from growing owing to the absence of sunlight and further oxygen diffusion from the air. The by-products of anaerobic digestion include hydrogen sulphide, which has a bad odour similar to rotten eggs, and methane gases. Owing to malodorous gas problems, anaerobic ponds were previously avoided. However, the odour problem can be minimized either by limiting the total weight of BOD loaded per unit volume per day or by recirculating facultative pond effluent to the anaerobic pond surface.
- *Aerated lagoons.* Aerated lagoons are a solution to the limitations of facultative and anaerobic lagoons. They are intended for cosmopolitan areas where land is limited and strict odour control is required. Oxygen is supplied to the process mechanically by means of diffusers or aerators. Owing to the shorter detention time, the biological community of an aerated lagoon is not as diverse as in an facultative lagoon.
- *Maturation ponds.* The coliform content of sewage could be very high. After the BOD has been degraded to levels acceptable for discharge to the receiving water body, the residual coliform content is still very high. Although the lagoon system is more efficient attremoving the coliform in the wastewater than the activated sludge or trickling filters, the effluent will require further treatment to remove the coliforms. This is done in a maturation pond.

This section will deal particularly with high-rate algal (as a particular type of aerobic lagoon) and maturation ponds.

1.5.1 High-rate algal ponds

The high-rate algal pond system seems to hold considerable potential. A high-rate algal pond can be considered as a continuous stirred tank reactor without cell recycle in which the growth rate is controlled by the hydraulic retention time (Grady and Lim 1980). In general, algal systems have been evaluated on the basis of total wastewater treatment, with less emphasis being placed on their potential for nutrient removal. In practice, the removal of nutrients by promoting algal growth on the effluent before discharge seems to be an attractive alternative (Hensman 1985).

1.5.1.1 Configuration of high-rate algal ponds

The efficiency of wastewater algal stabilization ponds depends on proper environmental factors, mainly light, temperature and the absence of toxic materials (e.g. chromate). Oxygen production in algal photosynthesis decreases with depth; for example, it is often five times greater near the surface than at a depth of 2.5 ft. Thus the use of shallow ponds of about 2.5 ft would be the most efficient, because below that depth the amount of oxygen used in algal respiration is at least as much as the oxygen released by photosynthesis. However, with an emphasis on factors such as temperature, water movement and problems of weed control, a pond 3–4 ft in depth would be desirable.

1.5.1.2 Principles of high-rate algal ponds

The transformations in a stabilization pond correspond closely to the natural purification that occurs in a stream receiving organic wastes. Algae present in the sewage pond, because of their photosynthetic activity, release oxygen into the water; the oxygen is then available to increase the aerobic decomposition of the organic wastes by bacteria. The rate of oxygen production by algae in daylight may be up to 20 times that of oxygen uptake by algae in respiration at night owing to bacterial and algal metabolism; the DO of the pond may reach about 25 mg/l in the daylight because of algal photosynthesis.

Phosphate removal by algae

Algae seem to offer the most easily exploited biological system for extracting phosphorus from domestic sewage. Laboratory and pilot-plant studies have indicated that in the presence of adequate amounts of light, soluble phosphate decreases equivalent to about 90% can be achieved with contact times as brief as 6–12 h.

In principle, algae can expedite phosphorus removal by two fundamentally different mechanisms: the metabolic assimilation of dissolved phosphorus to satisfy cell synthesis demand and the precipitation of insoluble phosphate compounds by a photosynthetically induced pH shift (Bogan *et al.* 1960).

Metabolic assimilation. The amount of phosphorus assimilated by algae for cellular synthesis depends on the intracellular phosphorus content of the organism and the productivity of the system. Because of a low productivity associated with photosynthetic processes, adequate phosphorus removal by assimilation mechanisms would require excessive hydraulic residence time and hence substantial land area.

Photosynthetically induced pH shift. The removal of phosphorus in excess of the amounts accounted for by metabolic assimilation is attributed to the change in calcium phosphate solubility as photosynthetic activity elevates the pH in the system. The induced pH shift is related to the algal consumption of inorganic carbon (CO_2) and an attendant shift in the carbonate–bicarbonate ion equilibrium.

To satisfy the carbon requirement for cellular synthesis, algae metabolize inorganic carbon from the aqueous medium (Goldman *et al.* 1972). In algal systems treating raw wastewater the inorganic carbon demand can be met primarily by the CO_2 generated by bacterial respiration. When CO_2 from alternate sources is insufficient to meet the demands for algal synthesis, the bicarbonate and carbonate components can supply free CO_2. Any increase or decrease in bicarbonate will result in an increase in the magnitude of the carbonate:bicarbonate ratio and consequently the pH. The increase in pH within the system leads to a change in the solubility of calcium phosphate among several other compounds.

1.5.1.3 Microbiology of high-rate algal ponds

Although many kinds of algae are sensitive to large amounts of organic wastes in their environment, others are tolerant and may be stimulated in their growth and reproduction by the presence of the wastes. So the sewage stabilization pond contains an algal population that is continually active and subject to substantial changes in number and constituent members. In general, all genera of algae recorded from sewage stabilization ponds can be classified as representatives of four groups: blue-green algae, diatoms, flagellate algae and green algae.

A study of sewage pond microbial populations in the U.S.A. and other American countries indicated a total of 125 genera. Among the most abundant were: *Chlorella*, *Ankistrodesmus*, *Scenedesmus*, *Euglena*, *Chlamydomonas*, *Oscillatoria*, *Micractinium*, *Golenkinia*, *Anacystis* and *Oocystis*.

In some ponds in Africa *Chlorella* was reported to be predominant in the first stage of treatment, but in the second stage its numbers fell off and *Spirulina* began to proliferate. In the third stage *Chlorella* had almost disappeared, and *Spirulina* predominated. Hensman (1985) for instance has developed on a pilot scale (i.e. a 260 m^2 algal basin, 400 mm deep) an algal growth system in which the biomass contained predominantly symmetrical tuft-like colonies of *Stigeoclonium* and diatoms. The relative proportions of *Stigeoclonium* and diatoms in the pilot plant were found to depend on the growth rate imposed on the system.

An examination of pond effluent for algae can give useful information. The predominance of *Chlorella* indicates that the pond is working at or over its capacity, whereas a mixed algal flora, with *Chlorella* being prominent only farther back in the pond, suggests that the pond could handle a higher load.

1.5.1.4 Application of high-rate algal ponds

A typical application of the high-rate algal pond is the Soetvelde waste treatment, Vereeniging in the Transvaal, South Africa. Soetvelde Farms Ltd at Vereeniging operates an integrated pilot-plant feedlot in which cattle are housed on concrete; manure is trampled into flushing channels from which it is flushed intermittently by streams of water (Cloete 1981). Solids are removed from the flush water by vibrating screens and the filtered slurry is passed through a high-rate algal pond into a fish pond from which the flush water is withdrawn. The purpose of the Soetvelde waste treatment system is first to stabilize organic matter (animal manure) to decrease its oxygen demand and to release plant nutrients, and then to grow micro-algae on the plant nutrients to produce oxygen for the stabilization process and to provide food for the growth of fish.

The storage pond has a surface area of 680 m^2, a depth of 0.35 m and a volume of 238 m^3; the high-rate algal pond has a surface area of 3,320 m^2, a depth of 0.45 m and a volume of 1,494 m^3; and the fish pond has a surface area of 12,000 m^2, a depth of 1.2 m and a volume of 14,400 m^3.

Periodically a specific volume of storage pond water is added to the high-rate algal pond. This pond is divided into six channels by means of asbestos partitions and produces algal and bacterial biomass, which is termed ABM. From the high-rate algal pond the water flows into a fish pond.

1.5.2 Maturation ponds

Maturation ponds are usually used after facultative and aerated lagoons and, in some instances, activated sludge and trickling filters to remove pathogenic organisms before the reuse of the treated wastewater, for instance for irrigation. A maturation pond refers to a lagoon up to 5 m deep in which the effluent from the facultative, anaerobic lagoon or other secondary treatment processes is treated to remove the coliform organisms.

Algal growth in the maturation ponds could take place rapidly. Thus in some instances the maturation ponds are also used as fish ponds. These fish are often used for animal feed, although in most cases the fish are fit for human consumption.

1.5.2.1 Configuration of maturation ponds

Maturation ponds are rather deep ponds, with depths ranging mostly from 3 to 5 m. As in anaerobic ponds, economic factors dictate the maximum pond depth, although in no case should the pond be made shallower than 3 m. This minimum value allows for siphoning of the effluent below the algal growth without disturbing the settled solids. As most of the algae in the maturation pond congregate in the upper 0.5 m where sunlight is available, the effluent is siphoned at 1.5–2.0 m depth to minimize the quantity of algae going to the effluent. In this way, only the motile algae are carried out because most of the non-motile algae stay in the top 0.4 m.

1.5.2.2 Principles of maturation ponds

The influent to the maturation pond contains a very low concentration of organic wastes: in general it may have filtered BOD values ranging from 20 to 60 mg/l only. However, the balance between oxygen supply and demand is not important in the design of maturation ponds. Hydraulic retention time is currently used as the major design criterion for maturation ponds.

Maturation ponds are operated in series of two or more ponds, each with a detention time of 5–10 days. The normal detention time in maturation ponds depends on the required bacterial removal efficiency, using faecal coliforms as an indicator. It is interesting to note that owing to the simplicity of the lagoon system, a higher microbial effluent quality is required than in most secondary treatment processes such as activated sludge or the trickling filter. First-order bacterial die-off kinetics is applied to maturation ponds.

Temperature effects. At 20 °C the coliform die-off constant (k_b) is 2.6/day. Given that coliform organisms are typical mesophiles, the decay rate should decrease with increasing temperature provided that the temperature does not exceed 40 °C. As the temperature approaches 37 °C, the coliform growth and reproduction rates reach their optimum. In fact in the presence of adequate food, the coliform population tends to increase rather than to decrease.

1.6 Conclusion

This chapter was intended to offer the reader an overview of biological methods for the treatment of wastewaters, their configurations, principles, microbiology, and applications. Although not exhaustive, it has the advantage of providing first-hand information for wastewater management. Indeed, the selection and design of an appropriate wastewater treatment process is a multidisciplinary approach: neither microbiologists nor engineers can go their own ways. Because no process should be considered a panacea, the selection of a suitable wastewater treatment method will include the following criteria:

- the economical feasibility of the process, i.e. the capital and running costs;
- the acceptability of the process in terms of environmental impacts, i.e. secondary effects such as odour and quantity of secondary waste;
- the effectiveness and reliability of the process, i.e. its ability to comply with the effluent requirements (standards);
- the organic loading of the influent, i.e. the type of wastewater to be treated;
- the land limitations, i.e. the land space available, site location (metropolis or rural area);
- technical factors such as expertise for maintenance.

1.7 References

Auling, G., Pils, F., Busse, H.J., Karrasch, S., Streichan, M. and Schön, G. (1991) Analysis of the phosphate accumulating microflora in phosphorus-eliminating, anaerobic–aerobic activated sludge systems by using diaminopropane as a biomarker for rapid estimation of *Acinetobacter* spp. *Appl. Environ. Microbiol.* **57** (12), 3585–3596.

Banks, C.J. and Walker, I. (1976) Sonication of activated sludge flocs and the recovery of their bacteria on solid media. *J. Gen. Microbiol.* **98**, 363–368.

Barnard, J.L. (1976) A review of biological phosphorus removal in the activated sludge process. *Water SA*. **2**, 136–144.

Blackall, L.L. (1993) Molecular identification of activated sludge foaming bacteria. In *First International Specialized Conference On Microorganisms in Activated Sludge and Biofilm Processes*, September, Paris, France, pp. 39–58.

Bogan, R.H., Albertson, O.E. and Pluntze, J.C. (1960) Use of algae in removing phosphorus from sewage. *J. Sanit. Eng. Div. ASCE* **86**, 1–20.

Bosch, M. (1992) Phosphorus uptake kinetics of *Acinetobacter* in activated mixed liquor. MSc. thesis, University of Pretoria, Pretoria, South Africa.

Boyd, R.F. (1984) *General microbiology*, pp. 414–418. Times Mirror/Mosby College, U.S.A.

Brodisch, K.E. (1985) Interaction of different groups of microorganisms in biological phosphate release. *Wat. Sci. Tech.* **17** (11/12), 139–146.

Brodisch, K.E.U. and Joyner, S.J. (1983) The role of microorganisms other than *Acinetobacter* in biological removal in activated sludge process. *Wat. Sci. Tech.* **15**, 117–125.

Buchan, L. (1980) The location and nature of accumulated phosphorus in activated sludge. D.Sc. thesis, University of Pretoria, Pretoria, South Africa.

Buchan, L. (1983) Possible biological mechanism of phosphorus removal. *Wat. Sci. Tech.* **15**, 87–103.

Buchan, L. (1984) Microbiological aspects. In *Theory, design and operation of nutrient removal activated sludge processes* (ed. H.N.S. Wiechers, G.A. Ekama, A. Gerber, G.F.P. Keay, W. Malan, G.V. Marais, D.W. Osborn, A.R. Pitman, D.J.J. Potgieter and W.A. Pretorius), pp. 9.1–9.6. Water Research Commission, Pretoria, South Africa.

Carberry, J.B. and Tenny, M.W. (1973) Luxury uptake of phosphate by activated sludge. *J. Wat. Pollut. Control Fed.* **45** (12), 2444–2462.

Capri, M.G. and Marais, G.v.R. (1975) pH adjustment in anaerobic digestion. *Wat. Res.* **9**, 307–313.

Cloete, T.E. (1981) Heterotrophic aspects and algal–bacterial interactions in a pond treatment system for cattle feedlot effluent. MSc thesis, UOFS, Bloemfontein, South Africa.

Cloete, T.E. and Steyn, P.L. (1987) A combined fluorescent antibody membrane filter technique for enumerating *Acineto-*

bacter in activated sludge. In *Biological phosphate removal from wastewaters. Advances in water pollution control* (ed. R. Ramadori), pp. 335–338. Pergamon Press, Oxford.

Cloete, T.E., Bosch, M. and Mienni, N.J.J. (1992) Organisms other than *Acinetobacter* capable of phosphorus removal from activated sludge mixed liquor. Paper presented at the European Conference on Nutrient Removal from Wastewaters, September, Leeds, U.K.

Cloete, T.E., Steyn, P.L. and Buchan, L. (1985) An autoecological study of *Acinetobacter* in activated sludge. *Environ. Technol. Lett.* **5**, 457–463.

Comeau, Y., Hall, K.J., Hancock, R.E.W. and Odhum, W.K. (1986) Biochemical model for enhanced biological phosphorus removal. *Wat. Res.* **20** (12), 1511–1521.

Deinema, M.H., Habets, L.H.A., Scholten, J., Turkstra, E. and Webers, H.A.A.M. (1980) The accumulation of polyphosphate in *Acinetobacter* spp. *FEMS Microbiol. Lett.* **9**, 275–279.

Dias, F.F. and Bhat, J.V. (1964) Microbial ecology of activated sludge. I. Dominant bacteria. *Appl. Microbiol.* **12**, 412–417.

Duchene, P.H. (1993) Biological foam: the cause–effect relationship, test results and combat strategy. In *First International Specialized Conference on Microorganisms in Activated Sludge and Biofilm Processes*, September, Paris, France, pp. 231–242.

Eckenfelder, W.W. Jr, Patoczka, J. and Watkin, A.T. (1985) Wastewater treatment. *Chem. Engr* **90**, 60–74.

Eikelboom, D.H. (1993) *Microthrix parvicella* puzzle. In *First International Specialized Conference on Microorganisms in Activated Sludge and Biofilm Processes*, September, Paris, France, pp. 267–275.

Ekama, G.A. and Marais, G.V.R. (1984) Nature of municipal wastewaters. In *Theory, design and operation of nutrient removal activated sludge processes* (ed. H.N.S. Wiechers, G.A. Ekama, A. Gerber, G.F.P. Keay, W. Malan, G.V. Marais, D.W. Osborn, A.R. Pitman, D.J.J. Potgieter and W.A. Pretorius), pp. 2.1–2.8. Water Research Commission, Pretoria, South Africa.

Ekama, G.A., Marais, G.V.R. and Siebritz, J.P. (1984) Biological excess phosphorus removal. In *Theory, design and operation of nutrient removal activated sludge processes* (ed. H.N.S. Wiechers, G.A. Ekama, A. Gerber, G.F.P. Keay, W. Malan, G.V. Marais, D.W. Osborn, A.R. Pitman, D.J.J. Potgieter and W.A. Pretorius), pp. 1–7. Water Research Commission, Pretoria, South Africa.

Fuhs, G.W. and Chen, M. (1975) Microbiological basis of phosphate removal in the activated sludge process for the treatment of wastewater. *Microb. Ecol.* **2**, 119–138.

Gerber, A., Mostert, E.S., Winter, C.T. and de Villiers, R.H. (1986) The effect of acetate and other short-chain carbon compounds on the kinetics of biological nutrient removal. *Water SA.* **12** (1), 7–12.

Gerber, A., de Villier, R.H., Mostert, E.S. and van Riet, C.J.J. (1987) The phenomenon of simultaneous phosphate uptake and release, and its importance in biological nutrient removal. In *Advances in Water Pollution Control* (A series of conferences on biological phosphate removal from wastewater) (ed. R. Ramadori), pp. 123–134. Pergamon Press, Oxford.

Gersberg, R.M. and Allen, D.W. (1984) Phosphorus uptake by *Klebsiella pneumoniae* and *Acinetobacter calcoaceticus. Proc. of IAWPRC conference*, Enhanced biolgical phosphorus removal from wastewater, Paris, France.

Goldman, J.C., Porcella, D.B., Middlebrooks, E.J. and Toerien, D.F. (1972) The effect of carbon on algal growth, its relationship to eutrophication. *Wat. Res.* **6**, 637–679.

Goncalves, R.F., Le Grand, L. and Rogalla, F. (1994) Biological phosphorus uptake in submerged biofilters with nitrogen removal. In *Proc. of IAWQ 17th Biennal International Conference*, 24–29 July, 1994, Budapest, Hungary, pp. 257–265.

Grady, C.P.L. and Lim, H.C. (1980) *Biological wastewater treatment – theory and applications*. Marcel Dekker, New York and Basel.

Harold, F.M. (1966) Inorganic polyphosphate in biology. Structure, metabolism and function. *Bact. Rev.* **30**, 772–794.

Hashimoto, A.C. and Chen, Y.R. (1979) The overall economics of anaerobic digestion. In *Anaerobic digestion* (ed. D.A. Stafford, B.I. Wheatley and, D.E. Hughes), pp. 449–466. Applied Science Publishers, London.

Hensman, L.C. (1985) The use of algae for the removal of phosphorus from secondary wastewater effluents. D.Eng. thesis, University of Pretoria, Pretoria, South Africa.

Hughes, D.E. (1979) What is anaerobic digestion? An overall view. In *Anaerobic digestion* (ed. D.A. Stafford, B.I. Wheatley and D.E. Hughes), pp. 1–13. Applied Science Publishers, London.

Iwai, S. and Kitao, T. (1994) *Wastewater treatment with microbial films*. Technomic Publishing Co., Lancaster, Basel.

Jenkins, D., Richard, M.G. and Diagger, G.T. (1986) *Manual on the causes and control of activated sludge bulking and foaming.* Water Research Commission, Pretoria, South Africa.

Kavanaugh, R. and Randall, C.W. (1993) Bacterial populations in a biological nutrient removal plant. In *First International Specialized Conference on Microorganisms in Activated Sludge and Biofilm Processes*, September, Paris, France, pp. 29–39.

Keay, G.F.P. (1984) Practical design consideration. In *Theory, design and operation of nutrient removal activated sludge processes* (ed. H.N.S. Wiechers, G.A. Ekama, A. Gerber, G.F.P. Keay, W. Malan, G.V. Marais, D.W. Osborn, A.R. Pitman, D.J.J. Potgieter and W.A. Pretorius), pp. 10.1–10.12. Water Research Commission, Pretoria, South Africa.

Lawson, E.N. and Tonhazy, N.E. (1980) Changes in morphology and phosphate uptake patterns of *Acinetobacter calcoaceticus* strains. *Water SA* **6** (3), 105–112.

Levin, G.V. and Shapiro, J. (1965) Metabolic uptake of phosphorus by wastewater organisms. *J. Wat. Pollut. Control Fed.* **37** (6), 800–821.

Li, A. and Sutton, P.M. (1983) Determination of alkalinity requirements for the anaerobic treatment process. Presented at 38th Industrial Waste Conference, Purdue University, Lafayette, Indiana.

Lötter, L.H. (1985) The role of bacterial phosphate metabolism in enhanced phosphorus removal from the activated sludge process. *Wat. Sci. Tech.* **17**, 127–138.

Lötter, L.H. and Murphy, M. (1985) The identification of heterotrophic bacteria in an activated sludge plant with particular reference to polyphosphate accumulation. *Water SA.* **11**, 179–184.

Marais, G.V.R., Loewenthal, R.E. and Siebritz, I.P. (1983) Observations supporting phosphate removal by biological excess uptake – a review. *Wat. Sci. Tech.* **15**, 15–41.

McCarty, P.L. and McKinney, R.E. (1961) Salt toxicity in anaerobic digestion. *J. Wat. Pollut. Control Fed.* **33**, 399–415.

McKinney, A.E. and Weichlein, R. (1953) Isolation of floc-producing bacteria from activated sludge. *Appl. Microbiol.* **1**, 259–261.

Mulder, J.W. and Rensink, J.H. (1987) Introduction of biological phosphorus removal to an activated sludge plant with practical limitations. In *Biological phosphate removal from wastewaters. Advances in water pollution control* (ed. R. Ramadori), pp. 213–223. Pergamon Press, Oxford.

Murphy, M. and Lötter, L.H. (1986) The effect of acetate on polyphosphate formation and degradation in activated sludge with particular reference to *Acinetobacter calcoaceticus*: a microscopic study. *Water SA.* **12** (2), 63–66.

Nagaoka, H. and Sugio, K. (1994) Effect of turbulenct structure on filament-type biofilm reactor. In *Proc. IAWQ 17th Biennal International Conference*, 24–29 July 1994, Budapest, Hungary, pp. 349–358.

Nicholls, H.A. and Osborn, D.W. (1979) Bacterial stress: prerequisite for biological removal of phosphorus. *J. Wat Pollut. Control Fed.* **51** (3), 557–569.

Nyns, E.J., Naveau, H.P., Chome, R. and Bertrand, Y. (1979) Digesters – a worldwide review. In *Anaerobic digestion* (ed. D.A. Stafford, B.I. Wheatley and D.E. Hughes), pp. 37–53. Applied Science Publishers, London.

Ohtake, H., Takahashi, K., Tsuzuki, Y. and Toda, K. (1985) Uptake and release of phosphate by a pure culture of *Acinetobacter calcoaceticus*. *Wat. Res.* **19** (12), 1587–1594.

Ouano, E.A.R. (1983) *Principles of wastewater treatment*, vol. 1 (*Biological processes*). National Science Development Board, Manila, Philippines.

Pauli, A. (1994) *The role of Acinetobacter sp. in biological phosphorus removal from forest industry wastewaters*. National Board of Waters and the Environment, Helsinki, Finland.

Pitman, A.R. (1984) Operation of biological nutrient removal plants. In *Theory, design and operation of nutrient removal activated sludge processes* (ed. H.N.S. Wiechers, G.A. Ekama, A. Gerber, G.F.P. Keay, W. Malan, G.V. Marais, D.W. Osborn, A.R. Pitman, D.J.J. Potgieter and W.A. Pretorius), pp. 11/1–11/16. Water Research Commission, Pretoria, South Africa.

Pfeffer, J.T. (1974) Temperature effects on anaerobic fermentation of domestic refuse. *Bioeng. Biotech.* **16**, 771–773.

Pfeffer, J.T. (1979) Anaerobic digestion process. In *Anaerobic digestion* (ed. D.A. Stafford, B.I. Wheatley and D.E. Hughes), pp. 15–33. Applied Science Publishers, London.

Pretorius, W.A. (1994) pH-controlled feed-on-demand for high-rate anaerobic systems. In *Proc. IAWQ 17th Biennal International Conference*, 24–29 July 1994, Budapest, Hungary, pp. 447–454.

Redman, J. (1987) Deep shaft treatment for sewage. *Chem. Engr*, Oct., pp. 12–13.

Shapiro, J. (1967) Induced rapid release and uptake of phosphate by microorganisms. *Science* **155**, 1269–1271.

Shapiro, J., Levin, G.V. and Zea, H.G. (1967) Anoxically induced release of phosphate in wastewater treatment. *J. Wat. Pollut. Control Fed.* **39** (11), 1810–1818.

Shoda, M., Oshumi, T. and Udaka, S. (1980) Screening for high phosphate accumulating bacteria. *Agric. Biol. Chem.* **44** (2), 319–324.

Slim, J.A. (1987) Some developments in the water industry in South Africa. *J. Wat. Pollut. Control Fed.* **86** (2), 262–271.

Stanier, R.Y., Ingraham, J.L., Wheelis, M.L. and Painter, P.R. (1988) *General microbiology*, 5th edn. Macmillan Education, London.

Streichan, M. Golecki, J.R. and Schön, G. (1990) Polyphosphate-accumulating bacteria from sewage plants with different processes for biological phosphorus removal. *FEMS Microbiol. Ecol.* **73**, 113–124.

Suresh, N., Warburg, R., Timmermam, M., Weels, J., Coccia, M., Roberts, M.F. and Halvorson, H.O. (1985) New strategies for the isolation of microorganisms responsible for phosphate accumulation. *Wat. Sci. Technol.* 17, 99–111.

Tam, N.F.Y., Wong, Y.S. and Lueng, G. (1992) Effect of exogeneous carbon sources on removal of inorganic nutrient by the nitrification denitrification process. *Wat. Res.* **9**, 1229–1236.

Tijhuis, L. (1994) The biofilm airlift suspension reactor – biofilm formation, detachment and heterogeneity. D.Eng. thesis, Delft University of Technology, Delft, The Netherlands.

Toerien, D.F., Gerber, A., Lötter, L.H. and Cloete, T.E. (1990) Enhanced biological phosphorus removal in activated sludge systems. *Adv. Microbiol. Ecol.* **11**, 173–230.

Van Haandel, A.C. (1994) Influence of the digested COD concentration on the alkalinity requirement in anaerobic digesters. In *Proc. IAWQ 17th Biennal International Conference*, 24–29 July 1994, Budapest, Hungary.

Van Loosdrecht, M.C.M., Tijhuis, L., Wiljdieks, A.M.S. and Heijnen, J.J. (1994) Population distribution in aerobic biofilms on small suspended particles. *Wat. Sci. Technol.* **29**, 377–384.

Van Strakenburg, W., Rensink, J.H. and Rijs, G.B.J. (1993) Biological P-removal: state of the art in the Netherlands. *Wat. Sci. Technol.* **27**, 317–328.

Venter, S.N., Lötter, L.H., de Haas, D.W. and MacDonald, L. (1989) The use of the analytical profile index in the identification of activated sludge bacteria: problems and solutions. *Water SA.* **15** (4), 265–267.

Wanner, J. (1993) The implementation of bulking control in designing activated sludge systems, In *First International Specialized Conference on Microorganisms in Activated Sludge and Biofilm Processes*, September, Paris, France, pp. 159–168.

Watanabe, Y., Okabe, S., Arata, T., and Haruta, Y. (1994) Study on the performance of an upflow aerated biofilter (UAB) in municipal wastewater treatment. In *Proc. IAWQ 17th Biennal International Conference*, 24–29 July 1994, Budapest, Hungary, pp. 277–285.

Weibe, W.J. (1989) Phosphorus, sulfur, and nitrogen cycles in mangrove forests. In *Recent advances in microbial ecology, ISME5*, pp. 312–317. Japan Scientific Societies Press.

Wentzel, M.C., Dold, P.L., Ekama, G.A. and Marais, G.V.R. (1985) Kinetics of biological phosphorus release. *Wat. Sci. Technol.* **17**, 57–71.

Young, J.C. and McCarty, P.L. (1969) The anaerobic filter for waste treatment. *J. Wat. Pollut. Control Fed.* **41**, 160–173.

Zhang, T.C. and Bishop, P.L. (1994) Experimental determination of the dissolved oxygen boundary layer and mass transfer resistance near the fluid–biofilm interface. In *Proc. IAWQ 17th Biennal International Conference*, 24–29 July 1994, Budapest, Hungary, pp. 297–306.

2. Biodiversity and microbial interactions in the biodegradation of organic compounds

Ronald M. Atlas

2.1 Biodiversity and microbial metabolism

Microorganisms have evolved the metabolic capacities of biodegrading most, if not all, naturally occurring organic substances (Alexander 1965). Some compounds, such as lignin and cellulose, are particularly resistant to microbial attack because of their limited water solubilities and complex chemical structures; only a limited diversity of microorganisms typically produce enzymes that attack such substances. Other simpler and water soluble compounds, such as glucose and proteins, are readily metabolized by numerous microorganisms, including many species of bacteria and fungi.

Despite the diversity of organic compounds that are subject to biodegradation and the vast array of different (biodiverse) microorganisms that metabolize them, there are only a few central metabolic pathways through which the metabolism of all organic compounds is channelled, namely glycolysis (most often the Embden–Myerhof pathway) and the tricarboxylic acid pathway. (Even here there is some diversity as some microorganisms have evolved variations in glycolysis such as the Entner–Doudoroff pathway.) The diversity of microbial metabolism of organic compounds occurs in large part in the various pathways that lead to these central metabolic pathways and to a smaller extent in the end products that are formed subsequent to these pathways (principally the different fermentation end products versus the carbon dioxide and water end products of respiration). To feed into the central metabolic pathways, various functional groups including amino groups must be removed because the intermediary metabolites of glycolysis and the tricarboxylic acid cycle are for the most part carboxylic acids. Diverse microorganisms have evolved different pathways for accomplishing the necessary metabolic steps that transform compounds into the intermediary metabolites of the Embden–Myerhof and tricarboxylic acid pathways.

Organic compounds that flow into waste treatment facilities and pollutants that enter the environment most commonly occur within mixtures. The initial steps in the metabolism of the various compounds in such mixtures are performed by diverse microorganisms, with the growth and metabolism of each type of microorganism being supported by one or more of the heterogenous compounds within the mixture. Even most synthetic compounds are sufficiently similar to naturally occurring compounds to be subject to microbial metabolism (Dagley 1975). A biologically diverse array of metabolic capacities has evolved that accomplishes vast biodegradation of numerous organic compounds in the environment and within waste treatment facilities. For example, numerous diverse microorganisms in activated sludge tanks of sewage treatment facilities are involved in the biodegradation of organic compounds. The biodegradation of these compounds within the wastes (decomposition) lowers the biochemical oxygen demand of the effluent water.

Although most organic compounds can be degraded, many are degraded only slowly so that they pass through waste treatment facilities without being biodegraded. It is possible, however, to design modern waste treatment facilities with sequential bioreactors so that the microorganisms within individual bioreactors can target the biodegradation of specific compounds. Great advances have been made in the design of bioreactors to remove specific compounds, both organics and inorganics. The necessary microbial populations and environmental conditions can be maintained within individual bioreactors to favour the degradation of specific compounds. For example, biological contactors with specific adherent biofilm microbial populations can be used to biodegrade azo dyes in wastestreams from textile manufacturers. The effluent waters from such treatment facilities are colourless unlike the brightly pigmented yellow waters released from many textile manufacturers lacking specialized bioreactors with specific microbial populations.

Many modern bioreactors used for waste treatment have carefully controlled zones with differing concentrations of oxygen. Different microbial populations flourish at various concentrations of molecular oxygen. Some microbial populations require molecular oxygen; others are killed by exposure to molecular oxygen. Anaerobic respiration is used in various waste treatment facilities to remove nitrate. Anaerobic conditions also favour the removal of chlorine and other halogen groups from organic compounds. Dehalogenation of organic compounds is thermodynamically favoured under anaerobic conditions; microorganisms have been found that produce dehalogenases and perform reductive dehalogenation (Suflita *et al.* 1982). Sulphate-reducing bacteria transform tetrachloroethylene to trichloroethylene and *cis*-1,2-dichloroethylene by anaerobic dehalogenation (Bagley and Gossett 1990). Trichloroethylene is also degraded aerobically by *Alcaligenes* by alternative pathways (Harker and Kim 1990).

2.1.1 Biodegradation of xenobiotic compounds

Not all compounds are biodegradable (Alexander 1965). Many xenobiotic compounds, which are chemicals synthesized by humans that have no close natural counterparts, have molecular structures and chemical bond sequences that are not recognized by existing degradative enzymes. These compounds resist biodegradation or are metabolized incompletely, with the result that some xenobiotic compounds accumulate in the environment. They pass through waste treatment facilities without modification or removal. Xenobiotic chemicals, which include pesticides, plastics and other synthetics, often persist in the environment indefinitely. The metabolic transport mechanisms and catabolic pathways for their biodegradation have not evolved (Alexander 1981).

Xenobiotic organic compounds might be recalcitrant (totally resistant) to biodegradation for a number of reasons (Alexander 1981). Unusual substitutions (such as with chlorine and other halogen atoms), unusual bonds or bond sequences (such as in tertiary and quaternary carbon atoms), highly condensed aromatic rings, and excessive molecular size (as in polyethylene and other plastics) are some of the common reasons for recalcitrance.

DDT, a once widely used pesticide, was one of the first molecules that was viewed as recalcitrant. DDT will pass through sewage treatment facilities without being modified. The persistence of DDT in the environment and its accumulation in high concentrations in tissues of animals raised the spectre that microorganisms fail to attack some compounds and that this leads to the environmental accumulation of those compounds.

Investigations have shown, however, that DDT actually can undergo various biochemical transformations; in other words, DDT is not recalcitrant in the absolute sense. The white rot fungus *Phanerochaete chrysosporium*, for example, is able to attack DDT (Bumpus and Aust 1987). An *Arthrobacter* species (with the addition of co-substrates, and a under regimen of alternating anaerobic and aerobic conditions) can mineralize DDT (convert the DDT to carbon dioxide, water and chloride ions) *in vitro* (Pfaender and Alexander 1972). Nevertheless, DDT mineralization in the environment either does not occur or occurs exceedingly slowly (Pfaender and Alexander 1972). Other than the unfavourable energetics of the dechlorination steps under aerobic conditions, no satisfactory explanation has been found for this fact.

Like DDT, trichlorophenoxyacetic acid (2,4,5-T) is relatively resistant to biodegradation and initially seemed to be recalcitrant. The resistance to biodegradation of 2,4,5-T is in marked contrast with the closely related compound dichlorophenoxyacetic acid (2,4-D), which is readily degraded by many aerobic soil microorganisms. The genes involved in the 2,4-D degradation pathway have been identified, and most have been cloned (Harker *et al.* 1989). Naturally occurring 2,4,5-T degraders have not been isolated. A strain of *Pseudomonas cepacia* that can degrade 2,4,5-T was isolated only after molecular breeding (a form of recombinant DNA technology – genetic engineering) in which a mixture of plasmid-bearing strains were maintained in a chemostat in the presence of 2,4,5-T (Kellogg *et al.* 1981; Chatterjee *et al.* 1982; Kilbane *et al.* 1982). Thus, although it can be degraded by a genetically engineered microorganism, 2,4,5-T will not be degraded by microorganisms in waste treatment facilities.

Complex PCB mixtures are also not attacked during their passage through sewage treatment facilities, even though a number of microorganisms have been isolated that transform PCBs (Chaudhry and Chapalamadugu 1991). Degradation of PCBs is performed by the white rot fungus *Phanerochaete* (Bumpus 1989), aerobically by *Acinetobacter* (Adriaens and Focht 1990) and *Alcaligenes* (Bedard et al. 1987), and anaerobically by reductive dehalogenation (Quensen *et al.* 1988). Extensive degradation of some PCB congeners has been found in soils and aquatic waters and sediments (Bedard *et al.* 1987; Brown *et al.* 1987; Novick and Alexander 1985). The specific congeners are differentially degraded, and various PCB products, according to their composition, exhibit different degrees of susceptibility to biodegradative transformations. Lower molecular mass congeners (those with fewer than four chloride substituents) are degraded under aerobic conditions; congeners of higher molecular mass (those with four or more chloride substituents) are reductively dehalogenated under anaerobic conditions but persist indefinitely under aerobic conditions.

In some cases, one portion of a molecule is

susceptible to degradation while the other is recalcitrant. Some acylanilide herbicides, for example, are cleaved by microbial amidases, and the aliphatic moiety of the molecule is mineralized (completely biodegraded). The aromatic moiety, being stabilized by chlorine substitutions, resists mineralization, but the reactive primary amino group can participate in various biochemical and chemical reactions leading to polymers and complexes that render the fate of such herbicide residues extremely complex. This has been shown for the metabolism of *N*-(3,4-dichlorophenyl)-propionamide (propanil; Figure 2.1) (Bartha and Pramer 1970). Microbial acylamidases cleave the propionate moiety, which is subsequently mineralized. A portion of the released 3,4-dichloroaniline (DCA) is acted upon by microbial oxidases and peroxidases, with the result that they dimerize and polymerize to form highly stable residues such as 3,3′,4,4′-tetrachloroazobenzene (TCAB) and related azo compounds.

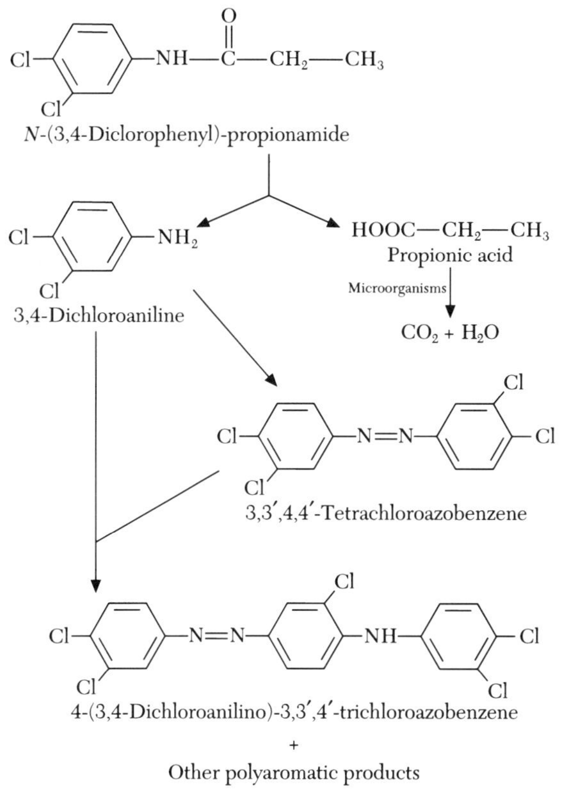

Figure 2.1. Microbial metabolism of the herbicide N-(3,4-dichlorophenyl)-propionamide (propanil). The aliphatic portion of the molecule is degraded, but the aromatic portion is dimerized and polymerized to persistent residues.

Thus a diverse array of chemical modifications to a xenobiotic compound can occur as a result of microbial metabolism. In some cases microbial attack leads to complete biodegradation (mineralization), at other times it results in minor modification (biotransformation) (for example, the removal of halogens or the addition of hydroxy groups), and at yet other times metabolic activities form more complex (higher molecular mass) compounds. Many of these reactions result in the detoxification of the compound and/or its immobilization (lowered solubility and bioavailability) so that there is less contact with living organisms that might be adversely affected.

2.1.2 Biodegradation of petroleum hydrocarbons

Petroleum hydrocarbons are among the most common environmental contaminants. Large oil spills attract massive media attention with the frequently shown pictures of dead seabirds. Most of the hydrocarbon contaminants, however, originate from much smaller spills and deliberate releases of oily wastes. Numerous underground storage tanks leak petrolum into soils and aquifers. Disposal of small quantities of waste oils and lubricants from automobiles frequently results in environmental contamination. Many of these waste oils enter sewer lines and pass through sewage treatment facilities; they are relatively undegraded by the microorganisms in those facilities unless the system has been specifically designed for the biodegradation of hydrocarbons as in those facilities used to treat oily ballast waters from oil tankers.

Susceptibility to biodegradation of the compounds in petroleum and refined oils varies with the type and size of the hydrocarbon molecule (Atlas 1984; National Research Council 1985). The most easily (rapidly) biodegraded compounds in petroleum generally are the n-alkanes of intermediate chain length (C_{10} to C_{24}). Higher molecular mass *n*-alkanes become increasingly resistant to biodegradation. Alkanes with molecular masses over 500 Da do not usually serve as carbon sources for microbial growth. Branching, in general, decreases the rate of biodegradation, as tertiary and quaternary carbon atoms interfere with degradation mechanisms or block degradation altogether.

The initial attack on alkanes occurs by enzymes that have a strict requirement for molecular oxygen, i.e. monooxygenases (mixed-function oxidases) or dioxygenases (Britton 1984; Singer and Finnerty 1984) (Figure 2.2). Most frequently the initial attack is directed at the terminal methyl group, forming a primary alcohol that, in turn, is further oxidized to an aldehyde and a fatty acid. Occasionally, both terminal methyl groups are oxidized in this manner, resulting in the formation of a dicarboxylic acid. This variation, described as diterminal or ω-oxidation, is one of several ways to bypass a block to β-oxidation due to branching of the carbon chain.

$$R\text{—}CH_2\text{—}CH_3 + O_2 + NADPH_2 \longrightarrow R\text{—}CH_2\text{—}CH_2\text{—}OH + NADP + H_2O$$

$$R\text{—}CH_2\text{—}CH_3 + O_2 \longrightarrow R\text{—}CH_2\text{—}CH_2\text{—}OOH$$

$$R\text{—}CH_2\text{—}CH_2\text{—}OOH + NADPH_2 \longrightarrow R\text{—}CH_2\text{—}CH_2\text{—}OH + NADP + H_2O$$

Figure 2.2. Pathway through which terminal oxidation of alkanes yields fatty acid moieties, which are metabolized further by β-oxidation.

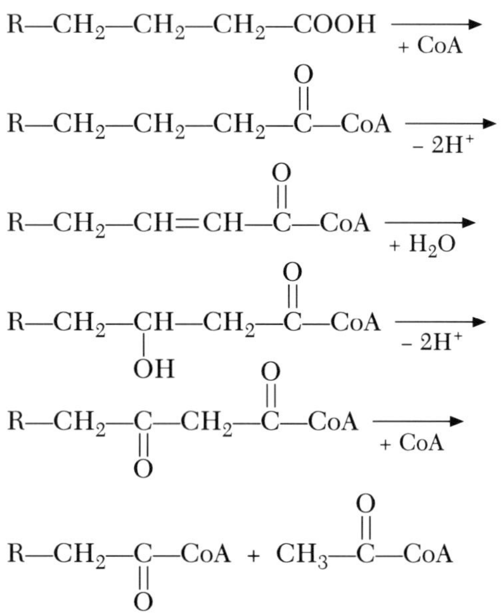

Figure 2.3. Pathway of β-oxidation results in the production of acetate from a carboxylic acid and a new carboxylic acid that is two carbon atoms shorter in chain length than the original substrate carboxylic acid.

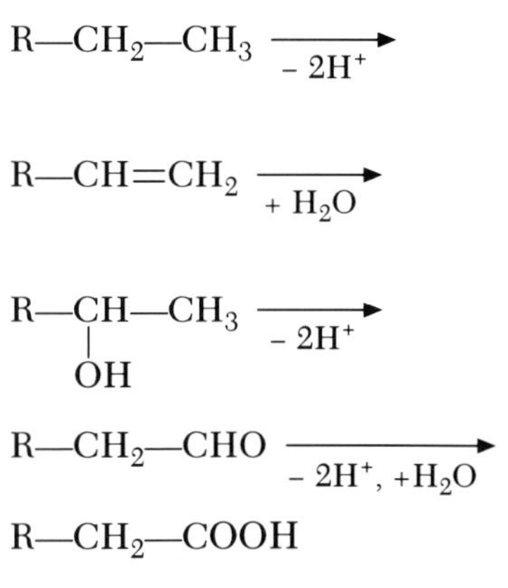

Figure 2.4. Pathway through which subterminal oxidation of alkanes yields fatty acid moieties, which are metabolized further by β-oxidation.

Once a fatty acid is formed, further catabolism occurs by the β-oxidation sequence (Figure 2.3). The long-chain fatty acid is converted to its acyl-coenzyme A form and is acted upon by a series of enzymes, with the result that an acetyl-CoA group is cleaved off and the fatty acid is shortened by a two-carbon unit. This sequence is then repeated. The acetyl-CoA units are converted to CO_2 through the tricarboxylic acid cycle. The end products of hydrocarbon mineralization are thus CO_2 and H_2O.

The β-oxidation sequence does not necessarily require the presence of molecular oxygen, and after the initial oxygenation, fatty acid biodegradation can proceed under anaerobic conditions. A direct dehydrogenation of an intact hydrocarbon – leading through a 1-alkene, an alcohol and an aldehyde to a fatty acid – can occur in the absence of molecular oxygen (Figure 2.4).

Some microorganisms attack alkanes subterminally; that is, oxygen is inserted on a carbon atom within the chain instead of at its end. In this way a secondary alcohol is formed first, which then is further oxidized to a ketone and finally to an ester. The ester bond is cleaved, yielding a primary alcohol and a fatty acid. The sum of the carbon atoms in the two fragments is equal to that of the parent hydrocarbon. The alcohol fragment is oxidized through the aldehyde to the fatty acid analogue, and both fragments are metabolized further by the β-oxidation sequence (Britton 1984; Singer and Finnerty 1984).

Alicyclic hydrocarbons without a terminal methyl group are biodegraded by a mechanism similar to subterminal oxidation. In cyclohexane (Figure 2.5), hydroxylation by a monooxygenase leads to an alicyclic alcohol. Dehydrogenation leads to the ketone. Further oxidation inserts an oxygen atom into the ring to form a lactone. The hydroxy group is oxidized, in sequence, to an aldehyde and carboxy group. The resulting dicarboxylic acid is further metabolized by β-oxidation. Microorganisms have been found that can grow on cyclohexane (Trower *et al.* 1985) and hence must be capable of performing the whole degradation sequence. More frequently, however, organisms capable of converting cyclohexane to cyclohexanone are unable to lactonize and open the ring, and vice versa.

Aromatic hydrocarbons are oxidized by dioxygenases to labile *cis,cis*-dihydrodiols that spontaneously convert to catechols (Figure 2.6). The dihydroxylated aromatic ring is opened by oxidative 'ortho cleavage', resulting in *cis,cis*-muconic acid. This is metabolized further to β-ketoadipic acid, which is oxidatively cleaved to the common tricarboxylic acid cycle intermediates succinic acid and acetyl-CoA. Alternatively, the catechol ring can be opened by meta cleavage, adjacent to rather than between the hydroxy groups, yielding 2-hydroxy-*cis,cis*-muconic semialdehyde. Further metabolism leads to formic acid, pyruvic acid and acetaldehyde.

Condensed aromatic ring structures, if degradable, are also attacked by dihydroxylation and the opening of one of the rings (Figure 2.7). The opened ring is degraded to pyruvic acid and CO_2, and a second ring is attacked in the same fashion. Many condensed polynuclear aromatic compounds, however, are degraded only with difficulty or not at all (Cerniglia 1984; Gibson and Subramanian 1984). One reason for resistance to biodegradation is that the induction of the enzymes responsible for degradation of polynuclear aromatic hydrocarbons depends on the presence of lower molecular mass aromatics (Heitkamp and Cerniglia 1988).

Importantly, eukaryotic microorganisms (fungi and algae) – like mammalian liver systems – produce *trans*-diols, whereas most bacteria oxidize aromatic hydrocarbons to *cis*-diols (Cerniglia 1984; Gibson and Subramanian 1984) (Figure 2.8). *Trans*-diols of various polynuclear aromatic hydrocarbons are carcinogenic, unlike *cis*-diols.

Most hydrocarbons are degraded only in the presence of oxygen. However, some low molecular mass aromatic compounds are metabolized under anaerobic methanogenic conditions (Grbic-Galic and Vogel 1987), with the oxygen for ring hydroxylation coming from H_2O (Vogel and Grbic-Galic 1986). Although the biodegradation of aromatics is definitely much slower under anaerobic than aerobic conditions, the rates of biodegradation of anaerobic toluene and *m*-xylene were considered significant enough for use in aquifer bioremediation (Zeyer *et al.* 1986, 1990).

2.2 Interactions between microbial populations

In addition to the variety of metabolic pathways for the biodegradation of numerous organic compounds exhibited by diverse species of microorganisms, there also are many interactions in microbial populations that influence the fates of compounds in the environment and within the bioreactors of waste treatment facilities. Microorganisms rarely exist in isolation; rather, numerous microbial populations of different types coexist. Often there is an interactive association between microorganisms, called a consortium, that results in combined metabolic activities. These microbial populations interact within a community. A variety of positive and negative population interactions lead to a stable functional community. The microbial community is structured so that each population contributes to its maintenance. Some microbial populations within the community adversely influence others and can even lead to their exclusion from the community. Other populations interactions are beneficial so that multiple microbial populations can live together at a particular location. A wastewater treatment facility, for example, will have a characteristic community involved in the degradation of wastes.

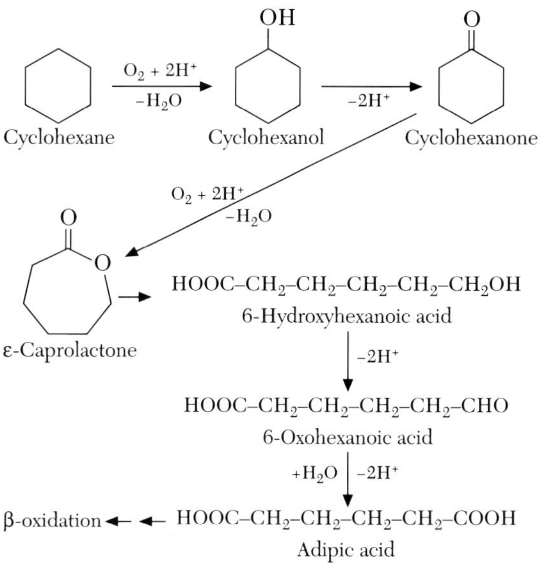

Figure 2.5. Microbial oxidation of cyclohexane as an example of metabolism of alicyclic hydrocarbons.

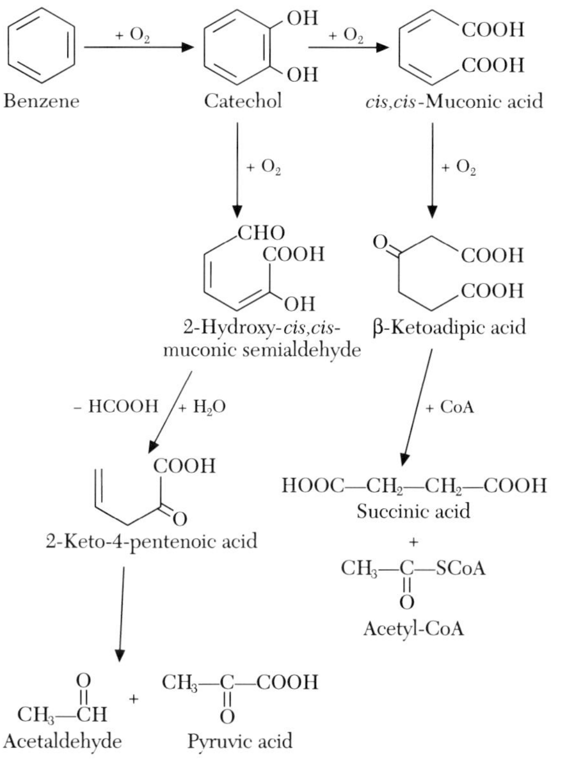

Figure 2.6. Microbial metabolism of the aromatic ring (simplified) by meta or ortho cleavage, as shown for benzene.

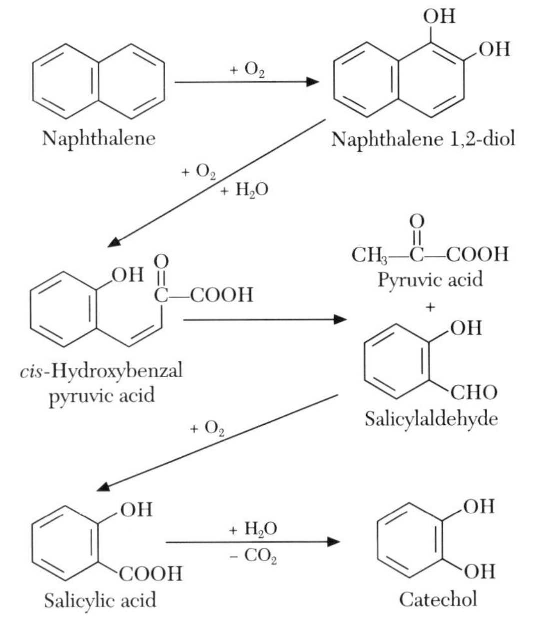

Figure 2.7. Microbial metabolism of a condensed aromatic ring structure (simplified) as shown for naphthalene. The resulting catechol is metabolized further by ortho or meta cleavage, as shown in Figure 2.6.

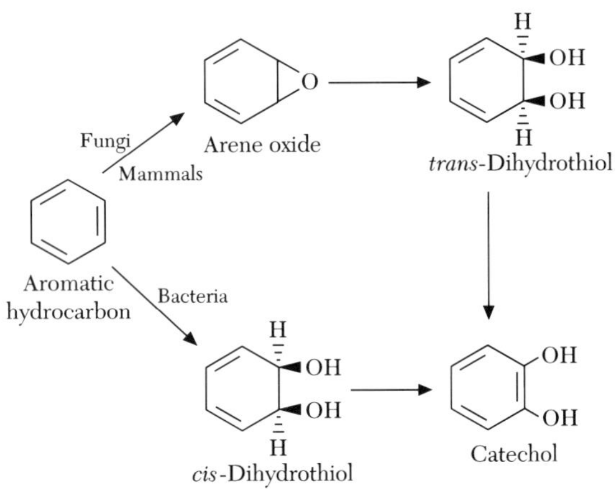

Figure 2.8. Fungi and other eukaryotic cells form trans-diols, whereas most bacteria form cis-diols when they oxidize aromatic hydrocarbons.

2.2.1 Commensalism and co-metabolism

One common type of population interaction, called a commensal relationship, occurs when one population benefits and the other remains unaffected. Commensalism is a unidirectional relationship between two populations; the unaffected population, by definition, does not benefit from, nor is it negatively affected by, the actions of the second population. The recipient population might need the benefit provided by the unaffected population but it might also be able to receive the necessary assistance from other populations with comparable metabolic capabilities.

Co-metabolism, in which an organism growing on a particular substrate gratuitously oxidizes a second substrate that it is unable to utilize as nutrient and energy source, is the basis for various commensal relationships. According to a strict definition of co-metabolism, the second substrate is not assimilated by the primary organism but the oxidation products are available for use by other microbial populations. Commensalism and co-metabolism play an important role in the biodegradation of alicyclic hydrocarbons (Perry 1984; Trudgill 1984). For example, *Mycobacterium vaccae* is able to co-metabolize cyclohexane while growing on propane; the cyclohexane is oxidized to cyclohexanol, which other bacterial populations can then utilize (Beam and Perry 1974). These bacterial populations benefit because they themselves are unable to metabolize intact cyclohexane; the *Mycobacterium* is unaffected because it does not assimilate the cyclohexane. Alicyclic hydrocarbons are often unable to serve as the sole carbon source for microbial growth unless they have a sufficiently long aliphatic side chain, but they can be degraded via co-metabolism by two or more cooperating microbial strains with complementary metabolic capabilities (Atlas 1981). Degradation of PCBs is typically by co-metabolism and is enhanced by the addition of less chlorinated analogues such as dichlorobiphenyl (Adriaens *et al.* 1989; Novick and Alexander 1985).

The activities of one microbial population can also make a compound available to another population without actually transforming the particular compound. For example, acids produced by one microbial population can release compounds that are bound or inaccessible to the second population. Such desorption processes are probably common in soil, where many compounds can be bonded to mineral particles or humic materials. Some organic molecules are converted by one population into substrates for other populations. Some fungi, for example, produce extracellular enzymes that convert complex polymeric compounds such as cellulose into simpler compounds such as glucose. The simpler compounds can be used by populations of other microorganisms that do not possess the enzymes needed to utilize the complex organic molecules. In some cases a competitive relationship develops for the available simpler substrates. In other cases the relationship can be truly commensal. Because the solubilized excess substrates would be lost to the first population by dilution in the environment, its use by a second population is not detrimental to the first one.

Another basis for commensalism between populations is the production of growth factors (Bell *et al.* 1974). Some microbial populations produce and excrete growth factors, such as vitamins and amino acids, that can be utilized by other microbial populations. Many bacteria, for example, depend on the production of vitamins by other microbial populations. As long as the growth factors are produced in excess and excreted from the neutral organism in the relationship, the two populations can have a commensal relationship.

Yet another basis for commensal relationships is the removal or neutralization of a toxic material. The ability to destroy toxic factors is widespread in microbial communities. The oxidation of hydrogen sulphide by *Beggiatoa* is an example of detoxification that benefits H_2S-sensitive aerobic microbial populations. *Beggiatoa* is not known to benefit from its relationship with the second population. The precipitation of heavy metals, such as mercury, by sulphate reducers provides an additional example of detoxification. The production of volatile mercuric compounds by bacterial populations in aquatic habitats removes this toxic metal from the habitat (Jeffries 1982). Some microbial populations are able to detoxify compounds by immobilization. As an example of such a commensal relationship, *Leptothrix* reduces manganese concentrations in some habitats, thereby permitting the growth of other microbial populations. Without the activities of the *Leptothrix* species the manganese concentrations in these habitats would be toxic to other microbial populations.

Commensalism also often results when the unaffected population, in the course of its normal growth and metabolism, modifies the habitat in such a way that another population benefits because the modified habitat is more suitable to its needs. For example, when a population of facultative anaerobes utilizes oxygen and lowers the oxygen tension, it creates a habitat suitable for the growth of obligate anaerobes. In such a habitat the obligate anaerobes benefit from the metabolic activities of the facultative organisms. The facultative organisms remain unaffected by the relationship as long as the two populations do not compete for the same substrates. The occurrence of obligate anaerobes within microenvironments of predominantly aerobic habitats is dependent on such commensal relationships.

2.2.2 Synergism and consortia

Synergism occurs when two populations cooperate so that each population benefits. Sometimes a consortium of two or more populations cooperate in this manner and supply each other's nutritional needs. Population 1 is able to metabolize compound A, forming compound B, but cannot go beyond this point without the cooperation of another population because it lacks the enzymes needed to bring about the next transformation in the pathway. Population 2 is unable to utilize compound A, but it can utilize compound B, forming compound C. Both populations 1 and 2 are able to perform the metabolic steps subsequent to the formation of compound C, producing needed energy and end products that neither population could produce alone. Such synergistic interactions are termed syntropism or cross-feeding.

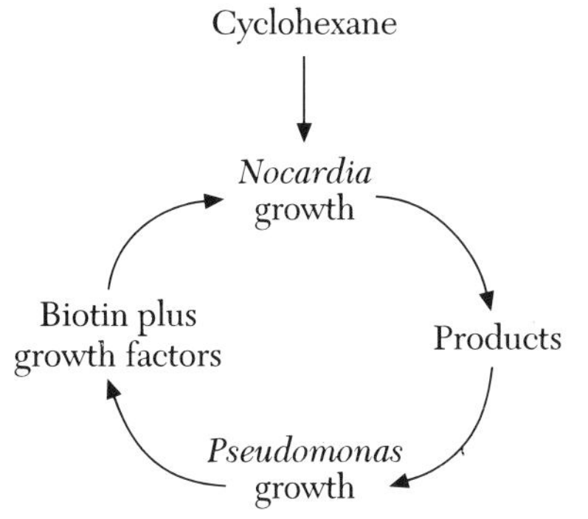

Figure 2.9. Synergistic degradation of cyclohexane by a Nocardia and a Pseudomonas. Nocardia supplies cyclohexane degradation products to Pseudomonas which supplies Nocardia with biotin. (Source: Slater 1978.)

Consortia of microorganisms can often accomplish through synergism the biodegradation of various organic compounds that cannot be biodegraded readily by a single microbial population. Cyclohexane can be degraded, for example, by a mixed population of a *Nocardia* species and a *Pseudomonas* species, but not by either population alone (Slater 1978) (Figure 2.9). The relationship is based on the ability of the *Nocardia* to metabolize cyclohexane, forming products that feed the *Pseudomonas* species. The *Pseudomonas* species produces biotin and growth factors required for the growth of the *Nocardia* species.

The fully chlorinated but unsaturated tetrachloroethylene ($Cl_2C{=}CCl_2$) can be dechlorinated by a consortium under anaerobic conditions in a stepwise fashion with partial conversion to CO_2. Tetrachloroethylene degradation has been demonstrated for a methanogenic bacterial consortium growing on acetate in an anaerobic reactor (Vogel and McCarty 1985; Galli and McCarty 1989). Chlorinated pesticides, including DDT, are likewise subject to reductive dehalogenation by consortia (Genthner *et al.* 1989a, 1989b). A nitrogen-heterocyclic herbicide, for example, has been shown to be attacked by reductive dehalogenation by a consortium of microorganisms acting synergistically (Adrian and Suflita 1990). Chlorobenzoates and chlorobenzenes are also degraded by this mechanism (Dolfing and Tiedje 1987; Fathepure *et al.* 1988; Stevens and Tiedje 1988; Stevens *et al.* 1988; Linkfield *et al.* 1989; Mohn and Tiedje 1990a, 1990b).

Aerobically, dichloromethane can serve as the only carbon source for a pseudomonad, and

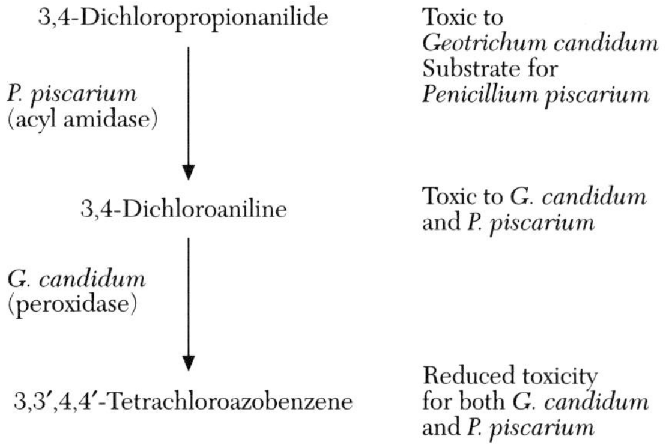

Figure 2.10. Synergistic degradation and detoxification of the herbicide 3,4-dichloropropionanilide (Propanil) by two soil fungi. (Source: Bordeleau and Bartha 1971.)

dichloroethane can serve as substrate for a microbial consortium (Stucki *et al.* 1981). Extensive aerobic degradation of trichloroethylene (TCE), a widely distributed halocarbon pollutant, by a methane-utilizing microbial consortium has been demonstrated (Fogel *et al.* 1986; Little *et al.* 1988). The low specificity of methane monooxygenase allows the conversion of TCE to TCE epoxide, which subsequently hydrolyses spontaneously to polar (formic acid, glyoxylic acid) products utilizable by microorganisms. *Methylococcus capsulatus* can convert chloro- and bromomethane to formaldehyde, dichloromethane to CO, and trichloromethane to CO_2 while growing on methane (Dalton and Stirling 1982). Thus, methanotrophic bacteria can biodegrade halocarbons and can be used within the bioreactors of waste treatment facilities to remove widespread halogen-containing pollutants, including TCE.

Several important degradation pathways of pesticides involve synergistic relationships. *Arthrobacter* and *Streptomyces* strains growing together are capable of completely degrading the organophosphate insecticide diazinon and can grow on this compound as the only source of carbon and energy (Gunner and Zuckerman 1968). This is accomplished by synergistic attack; alone neither culture can mineralize the pyrimidinyl ring of diazinon or grow on this compound. In a chemostat enrichment culture, *Pseudomonas stutzeri* is capable of cleaving the organophosphate insecticide parathion to *p*-nitrophenol and diethylthiophosphate, but is not capable of utilizing either of the resulting moieties (Munnecke and Hsieh 1976). *Pseudomonas aeruginosa* can mineralize the *p*-nitrophenol but is incapable of attacking intact parathion. Synergistically, the two-component enrichment degrades parathion with high efficiency, *P. stutzeri* apparently utilizing products excreted by *P. aeruginosa*.

More complex examples of synergism are based on the simultaneous removal of toxic factors and the production of usable substrates. Two soil fungi, *Penicillium piscarium* and *Geotrichum candidum*, are capable of the synergistic degradation and detoxification of the agricultural herbicide propanil (Bordeleau and Bartha 1971) (Figure 2.10). *Penicillium piscarium* cleaves propanil to propionic acid and 3,4-dichloroaniline. It uses propionic acid as a carbon and energy source, but is unable to process the toxic 3,4-dichloroaniline product any further. *Geotrichum candidum*, although unable to attack propanil, detoxifies 3,4-dichloroaniline by peroxidatic condensation to 3,3′,4,4′-tetrachloroazobenzene and other azo products. These end products are less toxic to these soil fungi than either 3,4-dichloroaniline or the parent herbicide, permitting, in the presence of additional carbon sources, increased growth yields for both fungal populations. Such synergistic growth stimulation occurs only in the presence of the herbicide. In the absence of the herbicide, the two fungi compete for available nutrient resources and consequently decrease each other's growth yield.

2.2.3 Competition

Often different microorganisms within complex communities attack different compounds but also, quite commonly, diverse microorganisms within those communities compete for the same substrates. Competition for available substrates is a main driving force that determines community structure and the diversity of biological populations that can coexist. In theory, competitive exclusion will eliminate all but one of the populations competing for an identical substrate. In some cases, however, microorganisms that utilize the same substrate coexist because of spatial separation within microhabitats or because of differing affinities for substrates at varying concentrations of that substrate that can lead to population fluctuations.

Differing environmental conditions will favour the competitive success of populations with adaptive features. Often a well-adapted population will become dominant, especially under stable environmental condtions. Greater diversity is favoured by unstable (fluctuating) environmental conditions. Under conditions of abundant substrate supply, rapidly growing microorganisms usually become the dominant poulations. This is true in sewage treatment facilities where the abundance of available organic substrates favours the growth of rapidly dividing bacteria such as coliforms and pseudomonads. These dominant microbial populations in sewage, which gain a competitive advantage because of their high intrinsic growth rates, are rapidly displaced in competition with other microbial populations of receiving waters as the concentration of organic compounds diminishes owing to decomposition and dilution; under lower nutrient conditions a more diverse community of slowly growing bacteria is favoured.

2.2.4 Predation

Some of the microorganisms involved in the decomposition of organic compounds within communities are themselves the substrates (prey) for other organisms. In activated sludge treatment facilities, for example, many of the bacterial populations that degrade the organic compounds in the waste are consumed by protozoa. Many diverse bacterial populations, including species of *Escherichia, Enterobacter, Pseudomonas, Achromobacter, Flavobacterium, Zoogloea, Micrococcus, Arthrobacter, Corynebacterium, Nocardia, Mycobacterium* and *Sphaerotilus*, grow in activated sludge on the heterogeneous organic substrates contained in sewage; these bacteria become the prey of protozoa and other higher organisms. Ciliate protozoa are especially important predators that graze on bacteria within waste treatment plant waters and sludges. These protozoa are in turn consumed by rotifers and other higher organisms to establish complex food webs. The degradative activities of the decomposers, primarily bacteria and fungi, form the base of these detrital food webs.

The bacteria and fungi in activated sludge tanks occur as free-living cells and within flocs. Slime-producing bacteria, especially *Zoogloea ramigera*, are especially important in forming flocs. Some of the dominant predatory ciliate protozoa, such as *Vorticella*, adhere to the floc and filter the surrounding water to consume suspended bacterial prey. Many of the bacteria within the floc are protected against these predatory protozoa and thus the floc represents a defence mechanism. Because the floc retains such a diverse community of bacteria capable of biodegrading the great array of compounds that occur in sewage, it is used as the inoculum for incoming raw sewage in the activated sludge treatment process; it also often is used for enrichment cultures when trying to isolate diverse microorganisms with novel biodegradative activities. The interacting microbial populations within activated sludge typify the resilience of complex microbial communities and the diversity of biodegradative activities that microbial populations are capable of performing.

2.3 Concluding remarks

Many diverse populations of microorganisms are involved in the biodegradation of organic compounds in waste treatment facilities and in the environment. These microbial populations exhibit a biologically diverse array of metabolic capacities so that they are able to decompose most but not all of the organic compounds in wastes and that contaminate the environment. The most resistant compounds to biodegradation are the xenobiotics and those with unusual chemical substitutions, such as chlorine substituents, and have limited water solubilities. Even many of these complex compounds are subject to biodegradation, albeit at slower rates and by only a limited (non-diverse) number of microorganisms. Besides the variety of metabolic pathways for the biodegradation of numerous organic compounds exhibited by diverse species of microorganisms, there also are many interactions between microbial populations that influence the fates of compounds in the environment and within the bioreactors of waste treatment facilities. Various strategies for cooperation have evolved within consortia, such as co-metabolism and syntrophism, that contribute to the biodegradation of heterogeneous organic compounds. Many of these interactions are complex and add to the diversity of metabolic activities in waste treatment facilities involved in biodegradation.

2.4 References

Adriaens, P. and Focht, D.D. (1990) Continuous coculture degradation of selected polychlorinated biphenyl congeners by *Acinetobacter* spp. in an aerobic reactor system. *Envir. Sci. Technol.* **24**, 1042–1049.

Adriaens, P.H., Kohler, P.E. and Kohler-Staub, D. (1989) Bacterial dehalogenation of chlorobenzoates and coculture biodegradation of 4,4-dichlorobiphenyl. *Appl. Envir. Microbiol.* **55**, 887–892.

Adrian, N.R. and Suflita, J.M. (1990) Reductive dehalogenation of a nitrogen heterocyclic herbicide in anoxic aquifer slurries. *Appl. Envir. Microbiol.* **56**, 292–294.

Alexander, M. (1965) Biodegradation: problems of molecular recalcitrance and microbial fallibility. *Adv. Appl. Microbiol.* **7**, 35–80.

Alexander, M. (1981) Biodegradation of chemicals of environmental concern. *Science* **211**, 132–138.

Atlas, R.M. (1981) Microbial degradation of petroleum hydrocarbons: an environmental perspective. *Microbiol. Rev.* **45**, 180–209.

Atlas, R.M. (ed.) (1984) *Petroleum microbiology*. Macmillan, New York.

Bagley, D.M. and Gossett, J.M. (1990) Tetrachloroethene transformation to trichloroethene and *cis*-1,2-dichloroethene by sulfate-reducing enrichment cultures. *Appl. Envir. Microbiol.* **56**, 2511–2516.

Bartha, R. and Pramer, D. (1970) Metabolism of acylanilide herbicides. *Adv. Appl. Microbiol.* **13**, 317–341.

Beam, H.W. and Perry, J.J. (1974) Microbial degradation of cycloparaffinic hydrocarbons via cometabolism and commensalism. *J. Gen. Microbiol.* **82**, 163–169.

Bedard, D.L., Wagner, R.E. and Brennan, M.J. (1987) Extensive degradation of arochlors and environmentally transformed polychlorinated biphenyls by *Alcaligenes eutrophus* H850. *App. Envir. Microbio.* **53**, 1094–1102.

Bell, W.H., Lang, J.M. and Mitchell, R. (1974) Selective stimulation of marine bacteria by algal extracellular products. *Limnol. Oceanogr.* **19**, 833–839.

Bordeleau, L.M. and Bartha, R. (1971) Ecology of a pesticide transformation: synergism of two soil fungi. *Soil Biol. Biochem.* **3**, 281–284.

Britton, L.N. (1984) Microbial degradation of aliphatic hydrocarbons. In: *Microbial degradation of organic compounds* (ed. D.T. Gibson), pp. 89–129. Marcel Dekker, New York.

Brown, J.F., Bedard, D.L. and Brennan, M.J. (1987) Polychlorinated biphenyl dechlorination in aquatic sediments (river sediments). *Science*, 236, 709–712.

Bumpus, J.A. (1989) Biodegradation of polycyclic aromatic hydrocarbons by *Phanerochaete chrysosporium*. *Appl. Envir.Microbiol.* **55**, 154–158.

Bumpus, J.A. and Aust, S.D. (1987) Biodegradation of DDT [1,1,1–trichloro–2,2–bis(4–chlorophenyl) ethane] by the white rot fungus *Phanerochaete chrysosporium*. *Appl. Envir. Microbiol.* **53**, 2001–2028.

Cerniglia, C.E. (1984) Microbial transformation of aromatic hydrocarbons. In *Petroleum microbiology* (ed. R.M. Atlas), pp. 99–128. Macmillan, New York.

Chatterjee, D.K., Kilbane, J.J. and Chakrabarty, A.M. (1982) Biodegradation of 2,4,5-trichlorophenoxyacetic acid in soil by a pure culture of *Pseudomonas cepacia*. *Appl. Envir. Microbiol.* **44**, 514–516.

Chaudhry, G.R. and Chapalamadugu, S. (1991) Biodegradation of halogenated organic compounds. *Microbiol. Rev.* **55**, 59–79.

Dagley, S. (1975) A biochemical approach to some problems of environmental pollution. In *Essays in biochemistry* (ed. P.N. Campbell and W.N. Aldridge), pp. 81–130. vol. 2, Academic Press, London.

Dalton, H. and Stirling, D.I. (1982) Co-metabolism. *Phil. Trans. R. Soc. Lond.* B **297**, 481–491.

Dolfing, J. and Tiedje, J.M. (1987) Growth yield increase linked to reductive dechlorination in a defined 3-chlorobenzoate degrading methanogenic coculture. *Arch. Microbiol.* **149**, 102–105.

Fathepure, B.Z., Tiedje, J.M. and Boyd, S.A. (1988) Reductive dechlorination of hexachlorobenzene to tri- and dichlorobenzenes in anaerobic sewage sludge. *Appl. Envir. Microbiol.* **54**, 327–330.

Fogel, M.M., Taddeo, A.R. and Fogel, S. (1986) Biodegradation of chlorinated ethanes by a methane-utilizing mixed culture. *Appl. Envir. Microbiol.* **51**, 720–724.

Galli, R. and McCarty, P.L. (1989) Biotransformation of 1,1,1-trichloroethane, trichloromethane, and tetrachloromethane by a *Clostridium* sp. *Appl. Envir. Microbiol.* **55**, 837–844.

Genthner, B.R.S., Price, W.A. and Pritchard, P.H. (1989a) Anaerobic degradation of chloroaromatic compounds in aquatic sediments under a variety of enrichment conditions. *Appl. Envir. Microbiol.* **55**, 1466–1471.

Genthner, B.R.S., Price, W.A. and Pritchard, P.H. (1989b) Characterization of anaerobic dechlorinating consortia derived from aquatic sediments. *Appl. Envir. Microbiol.* **55**, 1472–1476.

Gibson, D.T. and Subramanian, V. (1984) Microbial degradation of aromatic hydrocarbons. In *Microbial degradation of organic compounds* (ed. D.T. Gibson), pp. 181–252. Plenum Press, New York.

Grbic-Galic, D. and Vogel, T.M. (1987) Transformation of toluene and benzene by mixed methanogenic cultures. *Appl. Envir. Microbiol.* **53**, 254–260.

Gunner, H.B. and Zuckerman, B.M. (1968) Degradation of diazinon by synergistic microbial action. *Nature, Lond.* **217**, 1183–1184.

Harker, A.R. and Kim, Y. (1990) Trichloroethylene degradation by two independent aromatic-degrading pathways in *Alcaligenes eutrophus* JMP134. *Appl. Envir. Microbiol.* **56**, 1179–1181.

Harker, A.R., Olsen, R.H. and Seidler, R.J. (1989) Phenoxyacetic acid degradation by the 2,4–dichlorophenoxyacetic acid (TFD) pathway of plasmid pJP4: Mapping and characterization of the TFD regulatory gene, *tfd*R. *J. Bact.* **171**, 314–320.

Heitkamp, M.A. and Cerniglia, C.E. (1988) Mineralization of polycyclic aromatic hydrocarbons by a bacterium isolated from sediment below an oil field. *Appl. Envir. Microbiol.* **54**, 1612–1614.

Jeffries, T.W. (1982) The microbiology of mercury. *Prog. Indust. Microbiol.* **16**, 23–75.

Kellogg, S.T., Chatterjee, D.K. and Chakrabarty, A.M. (1981) Plasmid assisted molecular breeding – new technique for enhanced biodegradation of persistent toxic chemicals. *Science* **214**, 1133–1135.

Kilbane, J.J., Chatterjee, D.K., Karns, J.S., Kellog, S.T. and Chakrabarty, A.M. (1982) Biodegradation of 2,4,5-trichlorophenoxyacetic acid by a pure culture of *Pseudomonas cepacia*. *Appl. Envir. Microbiol.* **44**, 72–78.

Linkfield, T.G., Suflita, J.M. and Tiedje, J.M. (1989) Characterization of the acclimation period before anaerobic dehalogenation of halobenzoates. *Appl. Envir. Microbiol.* **55**, 2773–2778.

Little, C.D., Palumbo, A.V. and Herbes, S.E. (1988) Trichloroethylene biodegradation by a methane-oxidizing bacterium. *Appl. Envir. Microbiol.* **54**, 951–956.

Mohn, M.M. and Tiedje, J.M. (1990a) Catabolite thiosulfate disproportionation and carbon dioxide reduction in strain DCB-1, a reductively dechlorinating anaerobe. *J. Bact.* **172**, 2065–2070.

Mohn, M.M. and Tiedje, J.M. (1990b) Strain DCB-1 conserves energy for growth from reductive dechlorination coupled to formate oxidation. *Arch. Microbiol.* **153**, 267–271.

Munnecke, D.M. and Hsieh, D.P.H. (1976) Pathway of microbial metabolism of parathion. *Appl. Envir. Microbiol.* **31**, 63–69.

National Research Council (1985) *Oil in the sea: inputs, fates and effects*. National Academy Press, Washington, D.C.

Novick, N.J. and Alexander, M. (1985) Cometabolism of low concentrations of propachlor, alachlor, and cycloate in sewage and lake water. *Appl. Envir. Microbiol.* **49**, 737–743.

Perry, J.J. (1984) Microbial metabolism of cyclic alkanes. In *Petroleum microbiology* (ed. R.M. Atlas), pp. 61–97. Macmillan, New York.

Pfaender, F.K. and Alexander, M. (1972) Extensive microbial degradation of DDT *in vitro* and DDT metabolism by natural communities. *J. Agric. Fd Chem.* **20**, 842–846.

Quensen, J.F., Tiedje, J.M. and Boyd, S.A. (1988) Reductive dechlorination of polychlorinated biphenyls by anaerobic microorganisms from sediments. *Science* **242**, 752–754.

Sieburth, J.M. (1975) *Microbial seascapes*. University Park Press, Baltimore.

Singer, M.E. and Finnerty, W.R. (1984) Microbial metabolism of straight-chain and branched alkanes. In *Petroleum microbiology* (ed. R.M. Atlas), pp. 1–59. Macmillan, New York.

Slater, J.H. (1978) Microbial communities in the natural environment. In *The oil industry and microbial ecosystems* (ed. K.W.A. Chater and H.S. Somerville), pp. 137–154. Heyden and Sons, London.

Stevens, T.O. and Tiedje, J.M. (1988) Carbon dioxide fixation and mixotrophic metabolism by strain DCB-1, a dehalogenating anaerobic bacterium. *Appl. Envir. Microbiol.* **54**, 2944–2948.

Stevens, T.O., Linkfield, T.G. and Tiedje, J.M. (1988) Physiological characterization of strain DCB-1, a unique dehalogenating sulfidogenic bacterium. *Appl. Envir. Microbiol.* **54**, 2938–2943.

Stucki, G., Brunner, W., Staub, D. and Leisinger, T. (1981) Microbial degradation of chlorinated C1 and C2 hydrocarbons. In *Microbial degradation of xenobiotics and recalcitrant compounds* (ed. T. Leisinger, A.M. Cook., R. Hütter and J. Nüesch), pp. 131–137. Academic Press, London.

Suflita, J.M., Horowitz, A., Shelton, D.R. and Tiedje, J.M. (1982) Dehalogenation: a novel pathway for the anaerobic biodegradation of haloaromatic compounds. *Science* **214**, 1115–1117.

Trower, M.K., Buckland, R.M. , Higgins, R. and Griffin, M. (1985) Isolation and characterization of a cyclohexane-metabolizing *Xanthobacter* sp. *Appl. Envir. Microbiol.* **49**, 1282–1289.

Trudgill, P.W. (1984) Microbial degradation of the alicyclic ring. In *Microbial degradation of organic compounds* (ed. D.T. Gibson), pp. 131–180. Marcel Dekker, New York.

Vogel, T.M. and McCarty, P.L. (1985) Biotransformation of tetrachloroethylene to trichloroethylene, dichloroethylene, vinyl chloride and carbon dioxide under methanogenic conditions. *Appl. Envir. Microbiol.* **49**, 1080–1083.

Vogel, T.M. and Grbic-Galic, D. (1986) Incorporation of oxygen from water into toluene and benzene during anaerobic fermentative transformation. *Appl. Envir. Microbiol.* **52**, 200–202.

Zeyer, J., Eicher, P., Dolfing, J. and Schwarzenbach, P.R. (1990) Anaerobic degradation of aromatic hydrocarbons. In *Biotechnology and biodegradation* (ed. D. Kamely, A. Chakrabarty and G.S. Omenn), pp. 33–40. Gulf Publ. Co., Houston.

Zeyer, J., Kuhn, E.P. and Schwarzenbach, R.P. (1986) Rapid microbial mineralization of toluene and 1,3-dimethylbenzene in the absence of molecular oxygen. *Appl. Envir. Microbiol.* 52, 944–947.

3. Microbial population dynamics in biological wastewater treatment plants

Jiri Wanner

3.1 Definition and formulation of basic terms

The term 'population dynamics' was introduced quite recently in the research and practice of biological wastewater treatment. It was at the 14th IAWQ Biennial Conference in Brighton, U.K., in 1988 when the Specialist Group on Activated Sludge Population Dynamics (ASPD) was formally established. However, the inaugural meeting in Brighton followed several months of discussions among the 'core' members about the scope and aim of the Group, and **population dynamics** crystallized as the most suitable name for describing the subject of our interests. If 'dynamics' in its original meaning is defined as "a branch of any science concerned with forces that produce change in any field or system' (McLeod 1984), the term 'activated sludge population dynamics' can be understood as "a branch of water science and technology dealing with phenomena that govern the relationships between individual members of a complex microbial consortium traditionally called activated sludge". Although some biologists complain that this definition is too broad, wastewater microbiologists and engineers accepted this term quickly because it filled well the gap in wastewater treatment terminology.

The activated sludge plants designed at present for larger communities always include nitrogen and phosphorus removal simultaneously to the removal of organic compounds. Although sulphur compounds are not limited by the current effluent standards, sulphur metabolism can play a significant role in nutrient removal in activated sludge systems. Therefore it is necessary from the point of view of population dynamics to deal with biochemistry and microbiology of the following processes:

- removal of organic carbon
- reactions and importance of inorganic carbon
- reactions of different forms of nitrogen
- processes of biological removal of phosphorus
- metabolism of sulphur compounds.

These processes form rather complex biochemical systems. To describe the systems properly, and to avoid misunderstanding, it is necessary to use the same terms for describing the same processes. Unfortunately, no international unified terminology exists. Sometimes significant differences can be found between the terminologies used by microbiologists and by chemical engineers. Therefore the meaning of the terms used below should be clarified. The following definitions are based on terminology used in papers and reports of:

(1) IAWQ Task Group on Mathematical modelling for Design and Operation of Biological Wastewater Treatment,
(2) IAWQ Specialist Group on Nutrient Removal Processes from Wastewater,
(3) IAWQ Specialist Group on Activated Sludge Population Dynamics.

3.1.1 Substrate

Substrate is defined as the source of energy for living cells. Principally, there are three kinds of substrate in aquatic systems:

- *Light:* source of energy for **phototrophic** microorganisms
- *Inorganic compounds:* the energy is generated from the oxidation of reduced forms of such elements as N, S, Fe and Mn. Microbes obtaining energy in this way are called **chemolithotrophs (chemoautotrophs)**.
- *Organic compounds:* the energy is produced by biochemical oxidation of organic carbon to carbon dioxide. The microorganisms performing this reaction are called **chemoorganotrophs**.

Both the inorganic and the organic compounds serve as electron donors in biochemical oxidation–reduction reactions.

3.1.2 Carbon source

Besides energy, the living cells require a source of carbon for the synthesis of new biomass. Carbon can be metabolized in two forms: inorganic carbon or organic carbon.

3.1.2.1 Inorganic carbon

Inorganic carbon (dissolved CO_2, carbonate, bicarbonate) can be converted to organic cell materials by microbes called **autotrophs**. If the autotrophs use light (solar energy) as the substrate, they are known as **photoautotrophs**. In the case of using inorganic compounds as the substrate, the microbes are called **chemolithotrophs** (**chemoautotrophs**).

3.1.2.2 Organic carbon

Most aquatic microorganisms utilize organic carbon for the synthesis of new cells. Thus for the **chemoorganotrophs** the organic carbon represents both the substrate and the source of carbon for synthesis purposes. These microbes are often termed (**chemo**)**heterotrophs** in the literature.

The use of organic carbon as substrate and carbon source is energetically more advantageous for the microorganisms than chemolithotrophy. Thus in a complex environment, such as wastewaters, where both organic and inorganic substrates are present, the organotrophs will generally grow faster and more easily than the lithotrophs with a rather complicated mechanism of obtaining energy and forming new biomass. This is an important selective factor governing the composition of open mixed cultures such as activated sludge.

From the point of view of a microbial cell, the substrates and the carbon sources can be divided into two major groups:

- *External substrates and carbon sources*. These materials are present in the medium in which the cells are cultivated. Before they can be utilized by the cells, the materials should be transported from the environment to the cells' interior. This transport requires energy and is performed by transfer enzymes. The extracellular materials are either originally present in the cultivation medium (e.g. organic pollution in wastewaters) or synthesized by cells themselves (e.g. organic polymers). The metabolism when the cells use external substrates and carbon sources is termed **exogenous**.
- *Internal substrates and carbon sources*. A typical situation in the activated sludge process is when the external substrates are exhausted or present only in limiting concentrations. In such a case the cells do not cease their activities but switch the metabolism from exogenous to **endogenous**. During endogenous metabolism the materials accumulated and/or stored in the cells are metabolized. When these intracellular substrates are exhausted, the so-called autooxidation of cellular protoplasm may occur.

In connection with the terms 'substrate' and 'carbon source' the phenomenon of **mixotrophy** should be mentioned. Mixotrophic microorganisms can obtain energy by both organotrophic oxidation of organic compounds and chemolithotrophic oxidation of reduced inorganic compounds (especially sulphur compounds). The phenomenon of mixotrophy represents another selective factor in mixed cultures.

3.1.3 Nutrients

In general microbiology the term **nutrients** means all chemical elements that are utilized as building materials for cell synthesis. In wastewater treatment terminology only two elements are referred to as nutrients, namely nitrogen and phosphorus. The reason is that those two elements are considered to be the limiting nutrients for the growth of algae in eutrophic surface waters.

From the viewpoint of bacterial growth, elements such as nitrogen, phosphorus and sulphur are termed **macronutrients** because of their significant content in microbial biomass (in activated sludge 6–8% N and 2% P related to dry matter). The elements such as Fe, Ca, Mg, K, Mo, Zn and Co can be classified as **micronutrients**. The mass fraction of these elements in biomass is negligible but they are important in cells' metabolism as constituents of many enzymes. In cultivation conditions other than oxic (see the definition below), chemically bound oxygen can be considered a nutrient only.

3.1.4 Electron acceptors and cultivation conditions

The energy for maintaining cells' activities is generated from substrates by means of biochemical oxidation. The process of biooxidation means a transfer of electrons from oxidized substrates (**donors**), which are reduced in this process, to the compounds termed **acceptors**.

The modern terminology of cultivation conditions is based on different electron acceptors, which participate in individual biochemical reactions as described below.

3.1.4.1 Oxic conditions

In oxic conditions electrons from both organic and inorganic substrates are transferred to dissolved molecular oxygen, which is reduced and bound in a molecule of water. The most important biochemical processes in oxic conditions are as follows:

- oxic oxidation of organic compounds (organotrophic metabolism)
- chemolithotrophic oxidation of ammonia and nitrite to nitrate (nitrification) and of reduced sulphur compounds to sulphate
- synthesis of intracellular polyphosphate polymers (this process is not connected with an electron transport between donors and acceptors but it requires energy released from organotrophic metabolism of intracellular organic compounds).

3.1.4.2 Anoxic conditions

In anoxic conditions the electron acceptor

oxygen is replaced by nitrogen in oxidation state 5 (nitrate) or less frequently 3 (nitrite). By accepting five or three electrons respectively, nitrogen is reduced to oxidation state 0 (molecular nitrogen, N_2).

Two principal reactions are possible under anoxic conditions:

- anoxic organotrophic oxidation of organic compounds (when nitrate nitrogen as electron acceptor is reduced to nitrogen gas, the process is termed denitrification)
- anoxic chemolithotrophic oxidation of sulphide and elementary sulphur to sulphate.

The synthesis of intracellular polyphosphates in anoxic conditions seems to be possible, as the intracellular storage products can also be oxidized anoxically (Vlekke *et al.* 1988; Wanner *et al.* 1992).

3.1.4.3 Anaerobic conditions

Anaerobic conditions are characterized by the absence of both molecular oxygen and nitrate/nitrite nitrogen. The electrons in biochemical oxidation–reduction reactions can be transferred from organic substrates to:

(1) organic compounds (processes of fermentation, i.e. acido- and acetogenesis; methanogenesis is not considered in the activated sludge process)
(2) sulphate sulphur. In this dissimilatory sulphate reduction the sulphur is reduced by chemoorganotrophic microbes to elementary sulphur and to sulphide. Chemolithotrophic sulphate reduction is also possible but not very important in wastewater treatment.

The energy fixed in intracellular polyphosphates can be released under anaerobic conditions and used for the synthesis of organic intracellular storage materials from the products of anaerobic fermentation. In this reaction the oxidation state of organic carbon remains principally the same, so that no electron acceptors are needed.

The above defined individual cultivation conditions can be distinguished easily and exactly by measuring the oxidation–reduction potential (ORP). The values of ORP in oxic conditions are positive (higher than +50 mV as standard ORP), whereas ORP values in anaerobic conditions are negative (less than −50 mV). ORP values around 0 are typical of anoxic conditions.

3.1.5 Medium

A cultivation medium represents the real environment in which cells grow. The medium contains substrate, carbon source, nutrients, electron acceptors and other components. As well as its chemical composition, the medium is also characterized by parameters such as temperature, pH, ORP and osmotic pressure. In biological wastewater treatment a mixture of treated wastewater and mixed culture of activated sludge can be referred to as the **cultivation medium**. This medium is often called **mixed liquor**.

3.2 Microbial processes in biological wastewater treatment plants

3.2.1 Organic carbon removal

The metabolism of organic carbon is the most important process in biological wastewater treatment. The removal of organic compounds from wastewater after its contact with activated sludge can be divided into several steps:

- enmeshment of particles into the structure of activated sludge flocs
- entrapment and adsorption of colloidal materials
- sorption of soluble organic compounds of high molecular mass
- accumulation of single organic compounds with small molecules (readily biodegradable substrates) in bacterial cells.

The above processes are very fast and most organic substances are removed from the bulk liquid shortly after the contact of wastewater with the activated sludge. Eikelboom (1982) introduced the term **biosorption** to describe these simultaneous processes of initial, rapid organic substance removal. The extent of substrate removal by biosorption depends on the ratio of the mass of substrate in contact with the mass of activated sludge. Eikelboom (1982) named this ratio **floc loading**:

$$\text{floc loading} = \frac{(COD_i - COD_e)\,Q_i}{X_r Q_r} \qquad (3.1)$$

where

COD_i = influent COD
COD_e = effluent COD
Q_i = influent flow
Q_r = flow of return activated sludge
X_r = concentration of return activated sludge.

Biosorption also depends on the microbial composition of the particular activated sludge. The extent of biosorption is greater for activated sludges with good settling properties, i.e. activated sludges with a minor fraction of filamentous population. Eikelboom (1991) compared the biosorption of activated sludges originating from completely mixed systems with that from an activated sludge system with a so-called selector. Within the first 10 minutes of contact with wastewater, the activated sludge from the 'selector' system removed 85–92% of the total COD, whereas the activated sludge from the

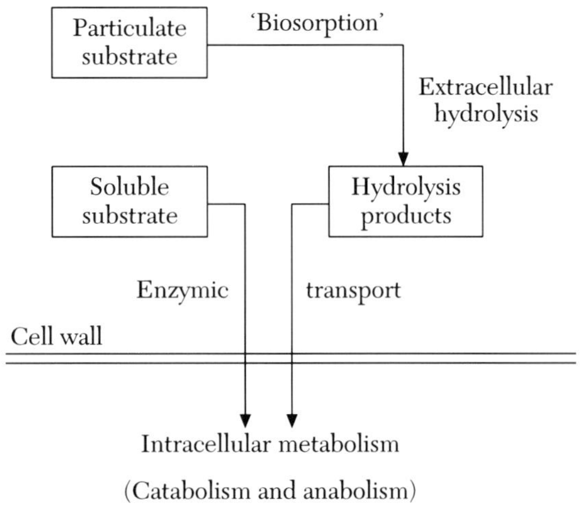

Figure 3.1. Schematic description of organic pollution removal from wastewaters by activated sludge microorganisms. (Redrawn from Wanner (1992).)

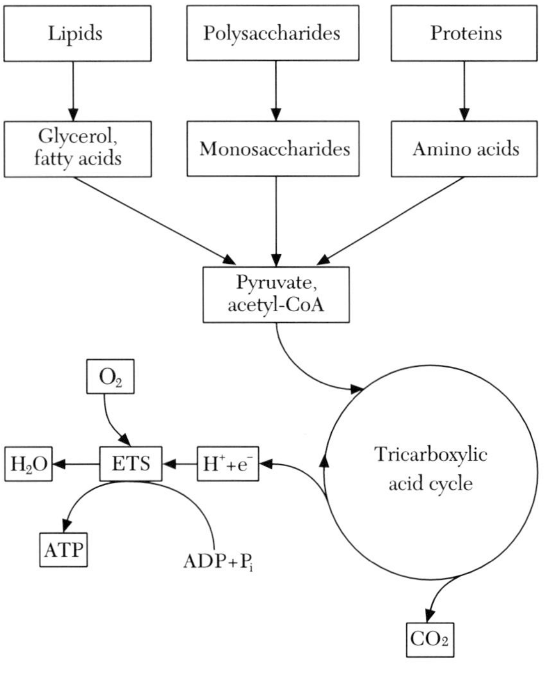

Figure 3.2. Schematic diagram of catabolism. (Redrawn from Wanner (1992).)

completely mixed system removed only 24–36%. These different capabilities of biosorption represent a selective factor that is used for the practical control of filamentous bulking.

Most of the organic substances sequestered by the activated sludge immediately after their contact with wastewaters (biosorption) are in a form that is not available for intracellular metabolism. The molecules of sorbed organic compounds are too large for them to permeate through the cell membranes. From the chemical point of view these substrates, which remain sorbed in the flocs of activated sludge and do not penetrate directly into the individual cells, are organic polymers. Polysaccharides, lipids and proteins are the most common high molecular mass compounds in wastewaters.

These polymers have to be degraded to structures with only few monomers in the chain or directly to monomers before their enzymic transport through the cell membrane. In biochemistry this process is termed **hydrolysis**; it is performed by specialized enzymes called **hydrolases**. The degradation of organic polymers to monomers forms the first part of organic carbon metabolism depicted in Figure 3.1.

After the extracellular hydrolysis of organic polymers, the fragments of polymers and single molecules are transferred to the cells, where they are metabolized by cells internal enzymic apparatus. The metabolism consists of two simultaneous processes:

- **Catabolism**, which generates energy from substrates for covering all energetic needs of cells
- **Anabolism** (assimilation) leading to the synthesis of new biomass.

The system of reactions that form catabolism is depicted schematically in Figure 3.2. Catabolism can be divided into three phases. The first phase completes the extracellular breakdown of complex molecules into simple low molecular mass compounds (monomers). The aim of the second phase of catabolism is to transform the numerous products of hydrolysis into few compounds entering the third, energy-producing phase.

The final part of catabolism is the respiration chain, in which the energy released from chemical bonds in the previous reactions is gained in a utilizable form. The respiration chain is a sequence of enzymic steps transferring the electrons and protons (hydrogen ions) to their final acceptors. Because of the transfer of electrons, this chain is also termed the **electron transport system** (ETS). In Figure 3.2 molecular oxygen serves as the final electron acceptor, but the electrons can also be transferred to other acceptors, for instance nitrate nitrogen under anoxic conditions. A comparison of the ETS under oxic and anoxic conditions is shown in Figure 3.3.

The energy generated in the ETS is not dissipated but conserved for a future use in the cell in te form of adenosine triphosphate (ATP). ATP is the most important and universal energy-transferring compound in living cells; its energy-poor counterpart is adenosine diphosphate (ADP).

The ability of the pair ATP + ADP to fix and release energy interconnects the mechanisms of energy production (catabolism) and energy consumption (anabolism). Energy in the form of ATP cannot be stored in the cells and must be converted into some energy 'cans' such as polysaccharides (both intracellular and extracellular), lipids and possibly proteins. In general, the ability of microbial cells to produce such

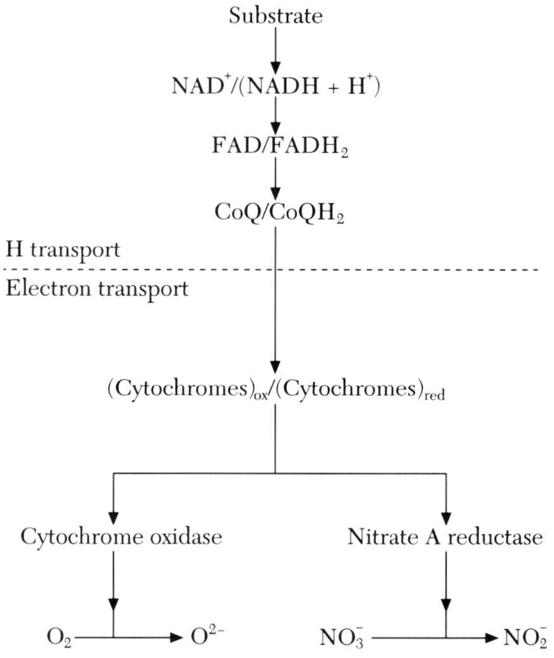

Figure 3.3. Electron transport systems under oxic and anoxic conditions. (Redrawn from Wanner (1992).)

'energy cans' is rather limited in comparison with animal cells. However, if some microbial species were to exhibit this ability, it would represent a great metabolic advantage in the competition with other microorganisms in mixed cultures.

The anabolic processes of biosynthesis follow essentially the same metabolic pathways as depicted in Figure 3.2 but in the reverse direction. The energy required for biosynthesis is derived from the catabolism and transferred by means of the pyridine nucleotide $NADP^+$ (nicotinamide–adenine dinucleotide phosphate) produced in the third phase of catabolism. The transfer of energy-rich activated hydrogen ($H^+ + e^-$) from catabolism to anabolism in the $NADP^+$ cycle is depicted schematically in Figure 3.4.

From the point of view of microbial selection in mixed cultures, the anabolic pathways leading to synthesis of storage product are of crucial importance. The organic storage products can store chemical energy from ATP for a long period and the energy stored inside the cells can even be transferred from one cultivation condition (reactor) to other one(s). The reason is that the solubility of organic storage products in water is very low, so that they do not exhibit any significant osmotic pressure. The consumption of energy for retaining these materials inside the cells is thus negligible. Typical storage products are polysaccharides but in contrast with plant cells, in which starch is the storage product, microbial cells produce either glycogen (a polysaccharide with the structure of a molecule similar to amylopectin) or extracellular heteropolysaccharides. The polysaccharides are generally synthesized from glucose, which is formed from products of the glyoxylic acid cycle. However, the organic storage product most frequently found in microbial cells is poly-β-hydroxybutyric acid (PHB). This polymer is synthesized by both aerobic and anaerobic organotrophic bacteria and by other microbes such as cyanobacteria. A simplified scheme of the synthesis of PHB from acetic acid (anabolism) and its degradation (catabolism) is shown in Figure 3.5 (Toerien *et al.* 1990).

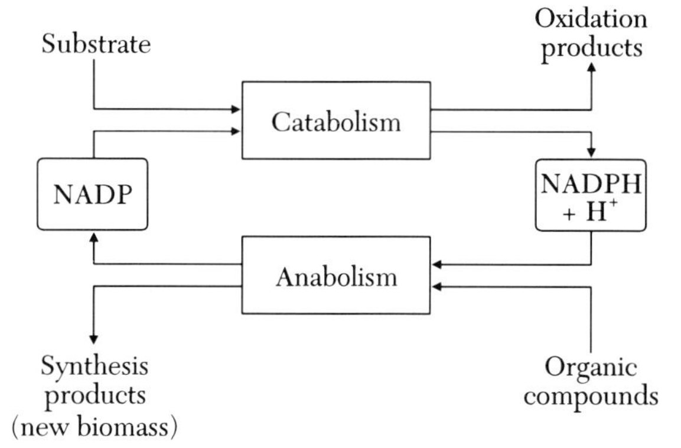

Figure 3.4. Schematic diagram of energy transfer in the $NADP^+$ cycle. (Redrawn from Wanner (1992).)

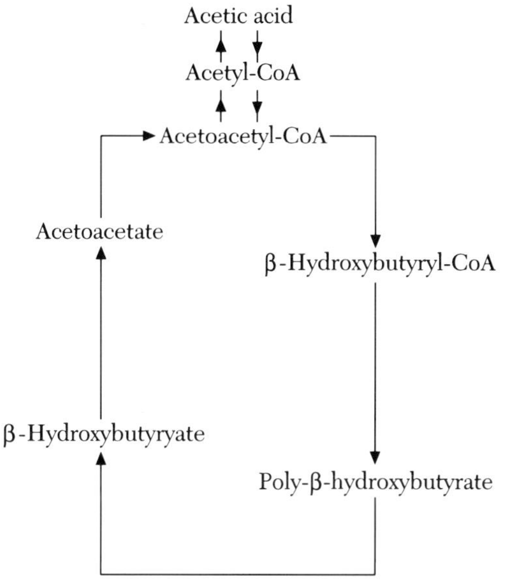

Figure 3.5. Simplified scheme of PHB synthesis and degradation. (Redrawn from Wanner (1992).)

According to Toerien *et al.* (1990), the catabolism of intracellular storage products is probably not initiated until all available exogenous substrates and carbon sources are exhausted. The catabolism of substances such as PHB is therefore connected with the endogenous metabolism. Both molecular oxygen and nitrate nitrogen can serve as electron acceptors in the final phase of the catabolism of intracellular organic storage products.

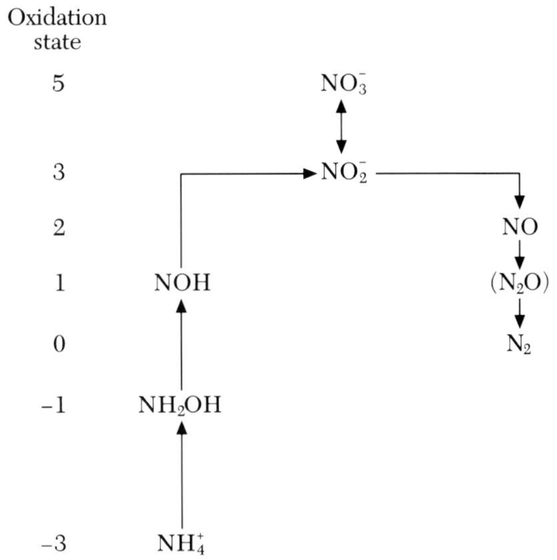

Figure 3.6. The changes in nitrogen oxidation state during nitrification and denitrification. (Redrawn from Wanner (1992).)

3.2.2 Nitrogen metabolism

3.2.2.1 Degradation of organic nitrogenous compounds

Proteinaceous organic substances are metabolized by the mechanism indicated in Figure 3.2. The biodegradable organic nitrogen from the amino group is converted to ammonia nitrogen by hydrolytic reactions, i.e. the oxidation state of nitrogen does not change in this process. As ammonia results from the degradation of organic nitrogenous compounds, the process is also called **ammonification**.

3.2.2.2 Nitrification

Nitrification is a typical example of a chemolithotrophic process when the microbes utilize energy generated from the oxidation of inorganic compounds. During this oxidation the oxidation state of nitrogen gradually changes from −3 to +5, as indicated in Figure 3.6. The first step is the oxidation of ammonium nitrogen to hydroxylamine, catalysed by monooxygenase. This reaction is **endergonic** (consuming energy); energy is produced only in the subsequent steps, i.e. in the oxidations of hydroxylamine and of nitrite. In the final step, the electrons are transferred from nitrite nitrogen to molecular oxygen via cytochrome a_1. The process of nitrification can formally be described by the following chemical equations:

$$NH_4^+ + 1.5O_2 \rightarrow NO_2^- + H_2O + 2H^+ + 250\text{ kJ} \qquad (3.2)$$

$$NO_2^- + 0.5O_2 \rightarrow NO_3^- + 75\text{ kJ} \qquad (3.3)$$

Stoichiometrically the oxygen requirement for reaction (3.2) is 3.43 g O_2/g N, and for the oxidation of nitrite (reaction 3.3) 1.14 g O_2/g N. The total specific consumption of oxygen for ammonia oxidation is 4.57 g O_2/g N. If we take into account that in simultaneous anabolic processes some electrons are transferred to carbon dioxide, which is thus fixed and converted to a new biomass, the specific consumption of oxygen for ammonia oxidation should be less than 4.57 g O_2/g N. The value of 4.2 g O_2/g N is most commonly found in the literature (Stensel and Barnard 1992).

The biochemical principles of chemolithotrophic nitrification in an aquatic environment as well as the microorganisms responsible are now well understood and have been described (Bock *et al.* 1988; Rheinheimer *et al.* 1988; Stensel and Barnard 1992). For our purposes it can be summarized that:

- nitrification is performed in two distinct processes by different kinds of chemolithotrophic microorganism
- most energy is produced and most of the oxygen is consumed in the first process
- chemolithotrophic nitrifying microorganisms are slow-growing and susceptible to various inhibitory effects including self-inhibition by substrate (un-ionized forms of nitrogen, i.e. NH_3 and HNO_2)
- under normal conditions the rate of oxidation of nitrite is higher than that of ammonia; thus nitrite does not accumulate in nitrifying activated sludge systems
- protons released during the oxidation of ammonia affect the buffer capacity; if the alkalinity is not high enough (at least 1.5–2 mmol/l at the end of nitrification), a significant decrease in pH can occur
- the specific consumption of oxygen for nitrification is rather high, which should be reflected in the design of aeration systems (dissolved oxygen concentration in the range 1.5–2.0 mg/l is recommended)
- although the microbiological literature refers to nitrifying microorganisms as strict 'aerobes', the process of nitrification can also be stabilized in activated sludges that are exposed for several hours to anoxic or anaerobic conditions; the enzymic apparatus of nitrifying bacteria is probably more complex than was thought in the past - for instance, the growth of nitrifying bacteria *Nitrobacter* in the absence of dissolved oxygen was experimentally verified in a pure culture (Bock *et al.* 1988).

3.2.2.3 Denitrification

As shown in Figure 3.6, the oxidation state of nitrogen can also be reduced biochemically. When reduced nitrogen is incorporated into newly synthesized biomass, the process is termed **assimilative nitrate reduction**. When nitrate nitrogen is reduced to elementary nitrogen and serves as an electron acceptor in the ETS (see Figure 3.3), the process is known as **denitrifi-**

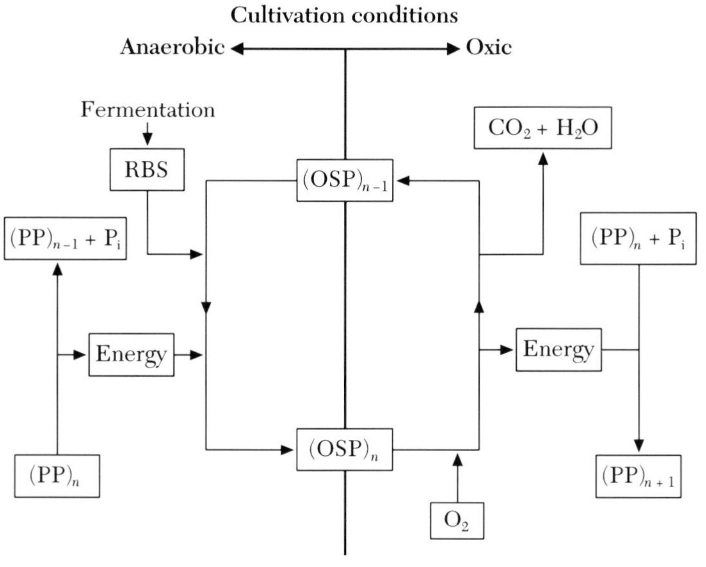

Figure 3.7. Transport of energy between anaerobic and oxic conditions. (Redrawn from Wanner (1992).)

cation. The process of denitrification can be described formally by the following equation:

$$8(H^+ + e^-) + H^+ + NO_3^- \rightarrow NH_4^+ + OH^- + 2H_2O \quad (3.4)$$

The donors of the activated hydrogen in Equation (3.4) are predominantly external readily biodegradable substrates, intracellular accumulated substrates or organic storage products (PHB). When all these substrates have been exhausted, slow endogenous anoxic respiration (denitrification) with $(H^+ + e^-)$ derived from cellular materials is also possible. As the rate of hydrolysis under anoxic conditions is slow, the availability of particulate substrates for denitrification is rather limited.

The specific yield of free energy released in the ETS under anoxic conditions is about 5% less than under oxic conditions when the final electron acceptor is O_2 (McCarty 1964). Ekama *et al.* (1979) compared the specific energy yield under oxic and anoxic conditions for different readily biodegradable substrates and for sewage. The difference did not exceed 7%. Thus denitrification can be considered to be an equivalent alternative to oxic respiration from both the metabolic and thermodynamic point of view. From the stoichiometry of oxic and anoxic respiration, it can be calculated that 1 g of NO_3^--N equals 2.86 g of O_2 in oxidation-reduction reactions. However, the actual consumption of readily biodegradable substrates for a full denitrification of 1 g of NO_3^--N expressed in COD units is estimated to be about 8.

In Equation (3.4) another important feature of denitrification can be seen: whereas protons are generated by nitrification, during denitrification protons are consumed. Thus in activated sludge systems with nitrification and denitrification the alkalinity can be partly replenished under anoxic conditions.

The process of denitrification is well understood from the point of view of its mechanism and kinetics (Schlegel 1985; Rheinheimer *et al.* 1988; Stensel and Barnard 1992). Nevertheless, there is still some uncertainty concerning the inhibitory effect of one denitrification intermetabolite, nitric oxide (NO), on denitrifying microorganisms (Casey *et al.* 1994).

3.2.3 Phosphorus metabolism

Phosphorus in the form of orthophosphate originates in municipal wastewaters from the degradation of phosphorus-containing organic substances and from the hydrolysis of the polyphosphates commonly used in commercial detergents. In conventional activated sludge systems, phosphorus from wastewaters is utilized only for the synthesis of new biomass components. The phosphorus content in activated sludges from conventional systems averages 2% in dry biomass. When the activated sludge is exposed to alternately anaerobic and oxic conditions, phosphorus is taken up by the cells in excess of synthesis purposes and the content of P in dry biomass can reach more than 10%. This phenomenon is called **enhanced biological phosphorus removal** (EBPR). The mechanism of EBPR represents an efficient way of transferring energy between anaerobic and oxic cultivation conditions. The transfer of energy by means of organic storage products (OSP) and polyphosphates (PP) is depicted schematically in Figure 3.7.

The phosphorus taken up in excess is stored in the cells in the form of polyphosphates counterbalanced with Ca^{2+}, Mg^{2+} and K^+ ions. The polyphosphate, together with lipidic and proteinaceous materials, forms intracellular granules called volutine after *Spirillum volutans*, a bacterium in which these granules were first observed. These granules are metachromatic

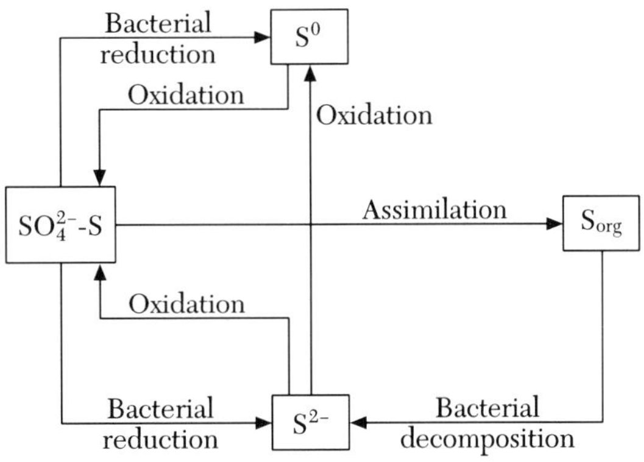

Figure 3.8. Simplified sulphur cycle (oxidation can be both photosynthetic and non-photosynthetic). (Redrawn from Wanner (1992).)

and change the colour of certain dyes, a property that is used for their identification. The primary function of the stored polyphosphate in most bacteria is that it serves as a phosphorus source for periods of phosphorus starvation. In some bacterial strains polyphosphate acts also as an energy source similar to the organic storage products (PHB, glycogen). The role of polyphosphate as intracellular energy storage material is decisive for the mechanism of EBPR as well as for the metabolic selection of bacterial species in activated sludge systems with anaerobic zone.

3.2.4 Sulphur metabolism

In an aquatic environment sulphur can be present in the following stable forms:

- reduced inorganic sulphur (oxidation state −2, sulphidic sulphur)
- organically bound sulphur (oxidation state −2, mostly in amino acids)
- elemental sulphur, S^0
- oxidized inorganic sulphur (oxidation state +6, sulphate sulphur).

Accordingly, the metabolism of sulphur consists of three basic processes which are to a large extent analogous to the reactions of nitrogen metabolism:

- mineralization of organic sulphur compounds (analogy: ammonification)
- oxidation of reduced inorganic sulphur compounds (analogy: nitrification)
- reduction of sulphate sulphur (analogy: denitrification).

A simplified sulphur cycle is shown in Figure 3.8.

3.3 Microbiology of activated sludge

Activated sludge should be understood as an artificial living ecosystem under the continuous influence of abiotic and biotic factors. Because of the necessity to reach rather low effluent concentrations of organic compounds (carbon and energy source) and inorganic nutrients, the activated sludge is cultivated under limiting conditions. This fact leads to a strong competition between individual groups of microorganisms: only the best adapted microbes win. As the influencing factors are not constant in wastewater treatment plants, the winners of the competition, dominating the activated sludge population, can change. Thus the microbial composition of activated sludges is not constant but reflects all the effects to which the activated sludge system is exposed.

Another characteristic feature of the mixed culture called activated sludge is that individual microbial cells are not separated in the cultivation medium but grow in aggregates. The ability of activated sludge microorganisms to flocculate is the most decisive property of activated sludges that enables us to use this mixed culture in large-scale installations. The flocculated aggregates (flocs of activated sludge) exhibit technologically acceptable sedimentation rates, and gravity sedimentation is the only economic way of separating the biomass from treated wastewaters in full-scale wastewater treatment plants. However, the ability to flocculate can be considered as a primary selective pressure in the mixed culture of activated sludge. The microorganisms that can clump or form flocs or at least that can be fixed into the flocs have the following selective advantages over freely growing cells:

- the microorganisms in flocs are retained in the activated sludge system, whereas the dispersed, freely growing cells are washed out
- the growth in flocs protects most microbial cells against predators.

From the microbiological point of view the activated sludge microorganisms can be divided into two major groups:

(1) decomposers, which are responsible for the biochemical degradation of polluting substances in wastewaters. This group is represented chiefly by bacteria, fungi and colourless cyanophyta. Osmotrophic protozoa can also ingest soluble organic substances but at the low concentrations of these materials in wastewater they cannot compete efficiently with bacteria.

(2) consumers, which utilize bacterial and other microbial cells as substrates. This group belongs to the so-called activated sludge microfauna and consists of phagotrophic protozoa and microscopic metazoa.

About 95% of the microbial population of activated sludges is formed by decomposers, especially by bacteria. This indicates that the role of microfauna in the removal of organic pollution and nutrients is only marginal.

3.3.1 Oxic organotrophic microorganisms

The genera *Bacillus, Pseudomonas, Alcaligenes, Moraxella* and *Flavobacterium* are reported to be able to degrade complex organic substrates by exo- and endoenzymes. Bacteria specialized to specific substrates can be concentrated in activated sludges after proper adaptation of the mixed culture to a given wastewater. Such specialized bacteria are, for example, *Proteus* spp. (proteinaceous materials) or *Achromobacter* spp. (lipids, acids and alcohols).

Besides organotrophic bacteria, microscopic fungi (micromycetes) and colourless cyanophyta (cyanobacteria) are involved in the oxic oxidation of organic substrates, especially of saccharidic and polysaccharidic compounds.

The organotrophic bacteria are both **floc-forming** and **filamentous**. *Zoogloea* spp. are considered typical floc-formers of activated sludges. About 30 different kinds of bacteria and cyanobacteria can grow in filaments in activated sludges (for more about filaments see below).

3.3.2 Fermentative bacteria

A fermentative conversion of organic compounds to volatile fatty acids is very important for the EBPR mechanism. In this connection, the presence of *Aeromonas punctata* and of the genera *Pasteurella* and *Alcaligenes* is stressed in the literature (see, for example, Brodisch and Joyner 1983; Brodisch 1985; Toerien *et al.* 1990).

3.3.3 Anoxic organotrophic microorganisms (denitrifiers)

The ability to use nitrate nitrogen as the final electron acceptor in biochemical reactions seems to be quite widespread among the activated sludge microorganisms. At least 40 species of aquatic microorganisms can denitrify. The genera *Achromobacter, Alcaligenes, Arthrobacter, Bacillus, Flavobacterium, Hypomicrobium, Moraxella* and *Pseudomonas* are typical organotrophic denitrifiers of activated sludges (Painter 1970; Rheinheimer *et al.* 1988; Stensel and Barnard 1992). Grabinska-Loniewska (1991) estimates that 82–97% of microorganisms in activated sludges from systems with an anoxic zone are able to denitrify. Activated sludge fungi perform only nitrate respiration, i.e. the first step of denitrification. The ability of filamentous microorganisms to utilize nitrate nitrogen as electron acceptor is crucial from the point of view of metabolic selection. In general, common filamentous microorganisms perform only the first step of denitrification, i.e. the reduction of nitrate to nitrite (Williams and Unz 1985a).

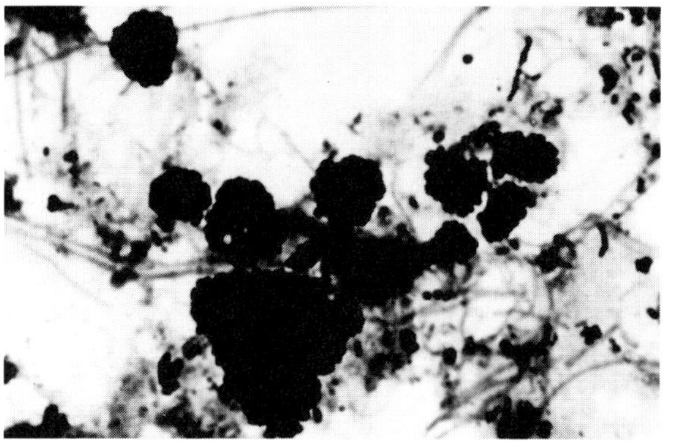

Figure 3.9. Photomicrograph of poly-P bacteria (clusters of black dots) (magnification ×630, Neisser staining).

3.3.4 Nitrifiers

The nitrifying bacteria were originally soil microbes. In an aquatic environment the following genera of nitrifiers are reported in the literature (Schlegel 1985; Rheinheimer *et al.* 1988; Stensel and Barnard 1992):

- *Nitrosomonas, Nitrosococcus, Nitrosospira* and *Nitrosocystis* for the oxidation of ammonia
- *Nitrobacter, Nitrospina* and *Nitrococcus* for the final oxidation of nitrite to nitrate.

In the activated sludge process the chemolithotrophic bacteria *Nitrosomonas* and *Nitrobacter* are considered the main nitrifiers. Heterotrophic nitrification can be attributed mainly to micromycetes, which are not important in activated sludge systems.

3.3.5 Polyphosphate-accumulating microorganisms

The ability to remove phosphate from wastewater by the EBPR mechanism is generally attributed to the genus *Acinetobacter (Acinetobacter calcoaceticus* var. *lwoffi),* which was identified in the isolates from EBPR activated sludges by means of fluorescent antibody techniques (Hart and Melmed 1982; Lötter and Murphy 1985; Cloete and Steyn 1988). The other microorganisms that might contain polyphosphate granules and thus contribute to the EBPR belong to the genera *Aeromonas, Arthrobacter, Klebsiella, Moraxella* and *Pseudomonas.* In the literature on the activated sludge process the polyphosphate-accumulating bacteria are generally referred to as **poly-P bacteria** (Figure 3.9). Some of the poly-P bacteria are seemingly able also to denitrify, i.e. to take up phosphate not only under oxic but also under anoxic conditions (Wanner *et al.* 1992; Kuba *et al.* 1994; Sorm *et al.* 1995).

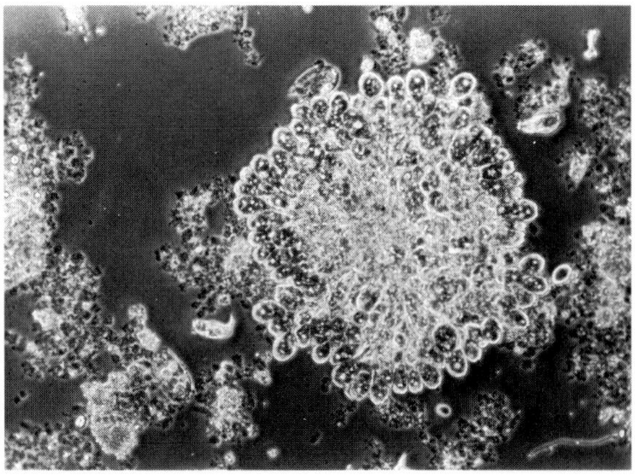

Figure 3.10. Photomicrograph of a large colony of attached ciliates (magnification ×75, phase contrast).

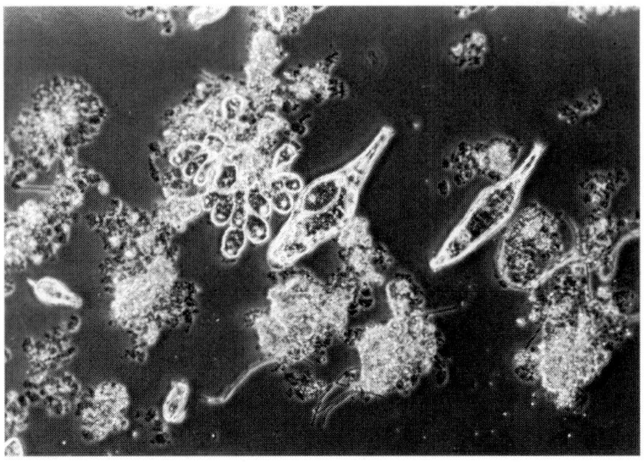

Figure 3.11. Photomicrograph of rotifers and ciliates (magnification ×75, phase contrast).

3.3.6 Sulphur bacteria

From numerous sulphur bacteria the colourless filamentous bacteria *Beggiatoa* and especially *Thiothrix* are important in the activated sludge process because they can cause bulking problems. *Thiothrix* seems to be a mixotrophic organism.

3.3.7 Microfauna of activated sludges

The microfauna of activated sludges consists of the following groups of protozoa and metazoa:

(1) Protozoa
- flagellates
- rhizopods
- ciliates - free swimming
  - crawling (grazing)
  - attached (stalked) (Figure 3.10)

(2) Metazoa
- nematodes
- rotifers (Figure 3.11)
- higher microfauna

Flagellates are the smallest protozoa found in activated sludges. Flagellates and rhizopods are mostly osmotrophic feeders, which means that they utilize soluble substrates like bacteria. Because the specific surface of protozoan cells is lower than that of bacterial cells, the efficiency of gaining substrate is less for osmotrophic protozoa than for bacteria. Thus the osmotrophic flagellates and rhizopods are typical only for start-up periods of the activated sludge process, before the stabilization of the bacterial population. Phagotrophic flagellates that use particulate substrates (for instance bacterial cells) follow the osmotrophic protozoa but are soon replaced by better organized ciliates in the development of activated sludge. Thus ciliates are characteristic protozoa of activated sludges cultivated under steady-state conditions. Free-swimming ciliates are connected with large numbers of dispersed bacteria in the bulk liquid. In contrast, grazing and attached ciliates, which require less energy because of their lower motility, can be found in well-flocculating activated sludges with limited numbers of dispersed bacteria (Buchan 1984). The attached ciliates can grow alone or in large colonies, which can even form whole flocs. Metazoa occupy a similar niche to that of attached ciliates, so that they are also found mostly in well-established activated sludges with good flocculation.

The protozoological literature attributes several important functions to microfauna in biological wastewater treatment, namely (Sudo and Aiba 1984):

- increased flocculation of bacteria
- removal of dispersed bacteria by adsorbing them on protozoan metabolites and by predation
- increase of the food/microorganism (F/M) ratio by reducing the number of bacteria as a result of predation
- direct uptake of substrates.

3.4 Biomass characterization

The biomass of the activated sludge is the active agent in biological wastewater treatment. To describe and control the processes occurring in wastewater treatment plants more precisely, the biomass should be characterized, at least from the following viewpoints:

(1) characterization and quantification according to metabolic activities
(2) identification and classification of microorganisms
(3) activated sludge quality (separability, dewaterability).

3.4.1 Metabolic activities

The most exact way to characterize the biomass of nutrient-removal activated sludge systems would be to quantify the presence of the above microorganisms or measure the activities of their specific enzymes. The methods of bacteria counting are based either (i) on culturing the

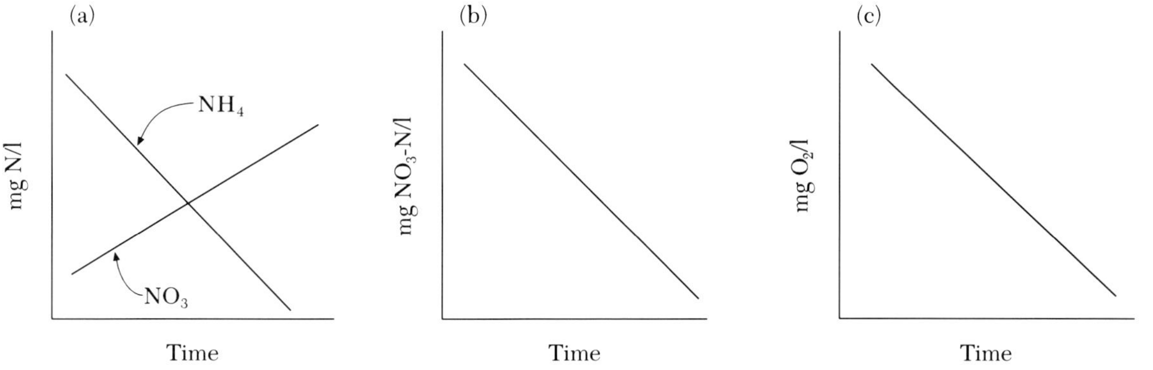

Figure 3.12. Schematic principles of AUR (a), NUR (b) and OUR (c) determinations (adapted from Kristensen et al. 1992).

sample of mixed culture under conditions favouring a defined metabolic group, enumerating the bacteria by means of Most Probable Number (MPN) method, surface plate method, or membrane filter technique, or (ii) on the isolation of particular microorganism and its quantification by means of specific techniques such as enzyme-linked immunosorbent assay and fluorescent antibody techniques. Although these methods are rather laborious, their applicability to the quantification of nitrifying bacteria (Sanden *et al.* 1994), poly-P bacteria (Lötter and Murphy 1985; Cloete and Steyn 1988; Kavanaugh and Randall 1994) and organotrophic and other bacteria (Schade and Lemmer 1994) has been proved recently in pilot-plant and full-scale nutrient-removal activated sludge systems. Techniques of measurement of specific enzyme activity (e.g. dehydrogenase activity in oxic conditions, nitrate reductase activity in anoxic conditions or polyphosphate kinase activity in EBPR) are also available but these tests are not very common in mixed bacterial cultures such as activated sludge.

The most common methods of characterization and quantification of activated sludge metabolic activities are indirect, i.e. the rate of consumption of components entering the biochemical reactions or the rate of formation of reaction products is measured instead of quantifying specific metabolic groups or enzymes (Wanner 1991). Figure 3.12 shows schematically the principles of determination of nitrification rate (measured as Ammonia Uptake Rate AUR or nitrate production rate), denitrification rate (Nitrate Uptake Rate, NUR) and oxic respiration (Oxygen Uptake Rate, OUR).

The EBPR process can be quantified by measuring the rate of orthophosphate release under anaerobic conditions and orthophosphate uptake under oxic/anoxic conditions. The Phosphorus Release Rate (PRR) under anaerobic conditions reflects to some extent the kinetics of anaerobic substrate uptake. However, PRR might be affected by other factors leading to the release of orthophosphate (e.g. pH) or by the simultaneous precipitation of released orthophosphate. Thus some authors suggest the quantification of not only the release of orthophosphate but also the counter-ions from intracellular polyphosphates, especially K^+ and Mg^{2+} (G.H. Kristensen, personal communication; Lindrea *et al.* 1993).

3.4.2 Identification and classification of microorganisms

The identification of activated sludge microorganisms is one of the most difficult tasks for wastewater microbiology. Fortunately there is no need in practice to identify all microorganisms in the mixed culture of activated sludge. For carbon removal microorganisms (both oxic and denitrifying) or nitrifiers, the present level of knowledge of these microbes is quite satisfactory from the practical point of view. However, there are still two groups of activated sludge microorganisms about which more precise information is needed:

- poly-P bacteria
- filamentous microorganisms.

3.4.2.1 Poly-P bacteria

The enhanced biological removal of phosphorus is associated mostly with the Gram-negative, Neisser-positive microorganisms from *Acinetobacter* species. Tandoi and Ramadori (1991) reviewed about 15 reports of the isolation and identification of *Acinetobacter* species during 1975–90. Tandoi and Ramadori concluded from this review that the key feature of poly-P bacteria, i.e. anaerobic organic carbon uptake, was not proved in pure cultures of isolates identified as *Acinetobacter*. This could mean either that the metabolic properties of *Acinetobacter* in mixed cultures differ from those in pure cultures or that other microbial species than *Acinetobacter* spp. are also responsible for EBPR. The latter possibility seems to be supported by studies done in South Africa and the U.S.A (Lötter and Murphy 1985; Cloete and

Steyn 1988; Kavanaugh and Randall 1994). In these studies *Acinetobacter* was not the dominant species in activated sludge with a well-established EBPR mechanism. However, direct proof that there are poly-P bacteria other than *Acinetobacter* is still lacking.

3.4.2.2 Filamentous microorganisms

Filamentous microorganisms are a specific morphological group of organotrophic (mostly) microorganisms growing in activated sludge. When present in excessive amounts, the filamentous microorganisms result in activated sludge separation problems known as bulking and foaming (see below). At present we distinguish about 30 different kinds of filamentous microorganism. Because of high demands on time and labour made by conventional microbiological identification procedures, the activated sludge filamentous microorganisms are not identified to taxa but to the so-called **types** introduced to wastewater treatment practice by Eikelboom (1975). With few exceptions (e.g. *Nocardia* spp., *Thiothrix* spp., *Sphaerotilus natans*), the taxonomic position of most Eikelbooms types is still unknown and the types are characterized by numbers (e.g. Type 0041 or Type 021N) or by taxonomically invalid names such as *Microthrix parvicella*.

The identification to types is based on some characteristic morphological features (cell and trichome size and shape, branching of trichomes) and on reactions to Gram and Neisser staining (Eikelboom and van Buijsen 1981; Jenkins *et al.* 1986). Unfortunately, the morphology and staining reactions can vary significantly depending on cultivation conditions and wastewater composition. For instance, an increased fraction of industrial wastes in treated wastewaters results in colours different from those described for the Gram stain in the above manuals. These problems cause there to be always some uncertainty in the results of microscopic identification. A typical example of this uncertainty is the most problematic filamentous microorganisms *Microthrix parvicella*: the scientists studying this filament cannot agree at present that all the laboratories worldwide are working with the same microorganism. Soddell *et al.* (1993) pleaded for more precise but rapid methods of identification that would be applicable in practice, and suggested gene probes for the identification and quantification of filamentous microorganisms *in situ*. Soon after this suggestion Blackall (1994) and Wagner *et al.* (1994) proved that this identification technique is feasible for activated sludge filamentous microorganisms. The molecular identification is based on specific sequences of ribosome 16 S and 23 S rRNA/DNA that can be detected by hybridization with fluoroscently labelled oligonucleotides. With this technique Blackall (1994) showed that the strains of foam-forming microorganisms *Nocardia amarae* and *Nocardia pinensis* most probably belong to the genus *Nocardia,* although there were some doubts in the literature about this allocation. Wagner *et al.* (1994) succeeded in identifying *Sphaerotilus natans*, *Haliscomenobacter hydrossis* and *Thiothrix nivea* by using an *in situ* hybridization technique with 16 S rRNA-targeted probes.

The improved knowledge of individual filamentous microorganisms will help to refine the existing system of classification of filaments. The classification of filamentous microorganisms can be done from different viewpoints. Eikelboom (1975) divided filamentous microorganisms in activated sludges into six groups on the basis of 'external' characteristic features that can be observed under the microscope: the shape and size of trichomes and individual cells, the presence of a sheath, the ability to form intracellular deposits, and staining characteristics (Gram, Neisser). Such a classification helps to clarify the seemingly chaotic realm of filamentous microorganisms. Jenkins *et al.* (1986) correlated the occurrence of individual filamentous types with technological parameters, operational conditions and wastewater characteristics. Five groups of filamentous microorganisms were established according to the causative conditions:

- low DO (dissolved oxygen)
- low F/M (low sludge loadings; high solids retention times)
- septic wastewaters (increased concentration of sulphide)
- nutrient deficiency (limiting concentrations of N,P, ?)
- low pH in mixed liquors.

Sometimes it is not easy to define exactly the range of the above causative conditions. For instance, the limiting concentration of O_2 depends on the actual floc loading by substrate (simultaneous diffusion and consumption in the floc interior). Is the solids retention time of 15 days high at winter temperatures (less than 10 °C)? Nevertheless, this classification showed indirectly the differences between Eikelbooms types in their kinetic properties and metabolic capabilities.

When nutrient-removal activated sludge systems were introduced in wastewater treatment practice, the diversity of types of filamentous microorganism in activated sludges increased. As was found by Blackbeard *et al.* (1988), Ziegler *et al.* (1988), Eikelboom (1991a,b), Foot (1992), K. Andreasen (personal communication) and by many others, the filamentous microorganisms dominating in activated sludges from nutrient removal systems were different from those reported earlier for conventional

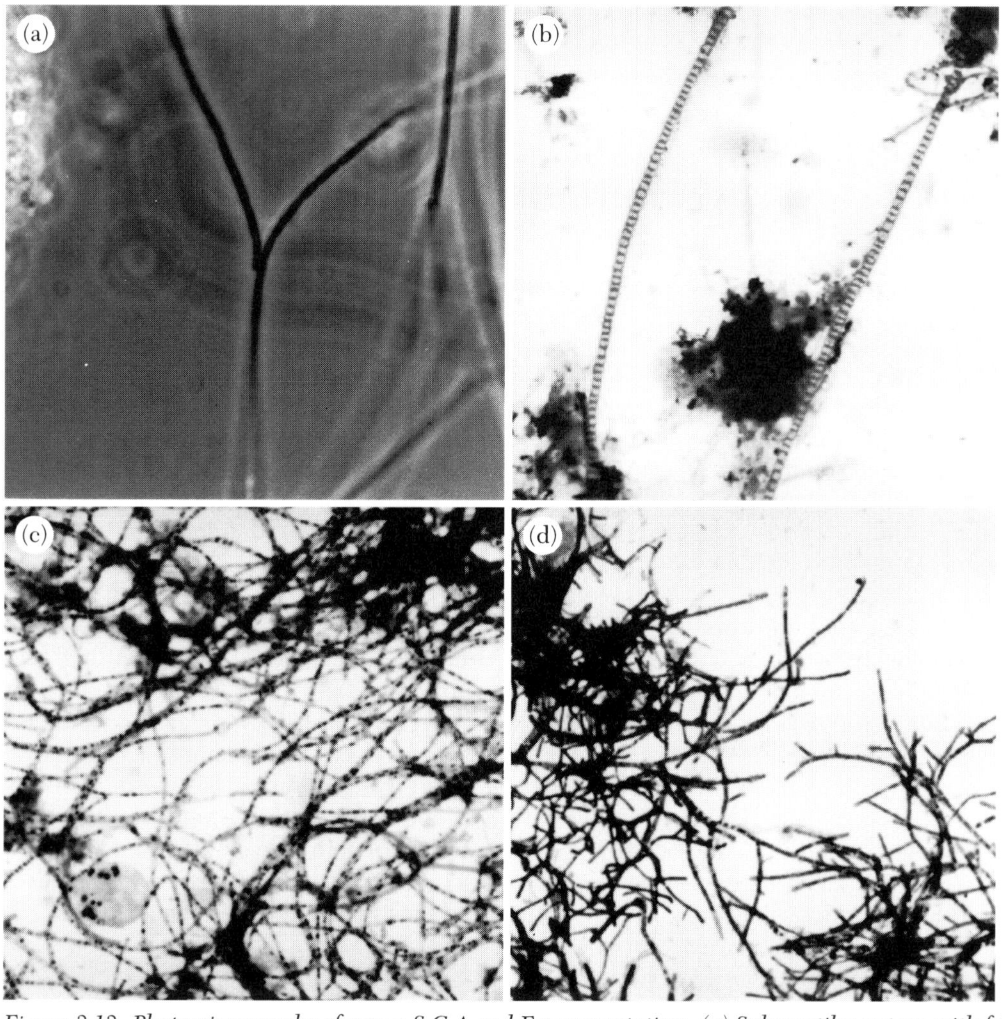

Figure 3.13. Photomicrographs of group S,C,A and F representatives: (a) Sphaerotilus natans with false branching (magnification ×750, phase contrast); (b) Cyanophyceae-like filaments (magnification ×750, Gram staining); (c) Microthrix parvicella (magnification ×750, Neisser staining); (d) nocardioform actinomycetes (magnification ×740, Gram staining).

(oxic) activated sludge systems. After the 'Milwaukee mystery' in 1969 (Anon. 1969), a new phenomenon of filamentous foaming has become typical of nutrient-removal activated sludge systems. It was logical to assume that the observed shift in filamentous population is connected with different properties of individual filamentous types.

Wanner and Grau (1989) summarized their own data and that from the literature and proposed the classification of filamentous microorganisms into four groups (Figure 3.13). The groups were set up on the basis of:

- morphological similarity (sheath, trichome, cells)
- similar staining reactions, intracellular deposits
- metabolic similarities, i.e. abilities to utilize substrates and gain energy under the same cultivation conditions
- occurrence in the same operational arrangements and conditions
- similarities in problems caused by the filamentous microorganisms.

Group S: Sphaerotilus-like oxic zone growers. Sheathed filamentous microorganisms that are able to utilize organic substrates only in oxic conditions. The presence of polyphosphates and PHB granules in cells can be observed but the rate of their formation and exploitation is not technologically important. Their occurrence in activated sludges is connected with saccharidic and other readily biodegradable wastewaters, higher sludge retention time (SRT), and low DO. Group S is comparable to Chiesa and Irvine's fast-growing, starvation-susceptible filaments (Chiesa and Irvine 1985). Characteristic representatives are *Sphaerotilus natans* and Type 1701, then Types 0041 and 0675.

Group C: Cyanophyceae-like oxic zone growers. Group C includes filamentous microorganisms morphologically resembling colourless blue-green algae: Type 021N and *Thiothrix*. Although not included in Eikelboom's manual, the genus *Leucothrix* should be added to group C. The taxonomic position of Type 021N is not yet clear; it seems that Type 021N is not identical with either the genus *Leucothrix* or the genus *Thiothrix* (Shimizu 1985; Williams and

Table 3.1. Response of filamentous microorganisms to systems combining kinetic selection (high concentration gradients) with metabolic selection under different cultivation conditions (Wanner 1992)

Group of filamentous microorganisms	Cultivation conditions Oxic	Anoxic	Anaerobic
S (e.g. *Sphaerotilus natans* and Type 1701)	–	–	–
C (e.g. Type 021N and *Thiothrix*)	–	–	–/+ (+ for sulphur filaments)
A (e.g. Microthrix parvicella, Type 0092, *Nostocoida limicola*)	0	0	0/+
F (e.g. nocardioform actinomycetes)	?	?/–	–/?

Key: –, suppression; + stimulation; 0, no effect; ?, effect uncertain.

Unz 1985b; Williams *et al.* 1987). Type 021N is a very common filamentous microorganism in bulking activated sludges from conventional activated sludge plants treating domestic or other wastewaters containing readily biodegradable substrates.

In contrast with oxic zone growers in group S, the microorganisms from group C (with the exception of *Leucothrix*) are able to metabolize sulphur. In particular, *Thiothrix* can take advantage of its mixotrophic way of life in systems with anaerobic zones.

Group A: all zones growers. The term 'all zones growers' means that these microorganisms are equipped with such a diverse enzymic apparatus that they can utilize substrates under all cultivation conditions in activated sludge plants. In addition, the all zones growers accumulate low molecular mass substrates under oxic and anoxic conditions and synthesize storage products with rates comparable to those of floc-formers.

Microthrix parvicella and Type 0092 exhibit undoubtedly the characteristic features of the all zones growers. *Nostocoida limicola* was commonly found in our pilots and one full-scale plant with anaerobic zones (Wanner and Grau 1989). Foot (1992) also added Type 0803 to this group.

Group F: Foam-forming microorganisms. The foam-formers are microorganisms that can produce biosurfactants enabling them to froth and create scum. The floating effect is supported by hydrophobic surfaces of cells. The hydrophobic cells stabilize the air bubbles or oil droplets entrapped in the foam.

The formation of biological foam is mostly connected with actinomycetes, namely the so-called nocardioform actinomycetes (*Nocardia, Rhodococcus*) or bacteria such as *Microthrix parvicella*. These microorganisms can be considered as primary foam-formers together with Type 0092, Type 0041 and *Nostocoida limicola*. Other filamentous microorganisms commonly found in biological foams are probably only trapped in already established scums.

The differences between individual groups S, C, A and F can be illustrated by the response of filamentous microorganisms to systems combining kinetic selection (i.e. systems with high concentration gradients) with metabolic selection under different cultivation conditions, as shown in Table 3.1 (for the selection mechanisms see Section 3.5).

3.4.3 Activated sludge quality

The control of activated sludge settling and dewatering properties is the most important task of activated sludge population dynamics. Even the most sophisticated nutrient-removal activated sludge systems cannot work properly without good settling, thickening and dewatering properties of activated sludge. The activated sludge separation problems have several causes but the following two are principal:

- excessive growth of filamentous microorganisms (bulking and foaming)
- poor formation of activated sludge flocs.

3.4.3.1 Measurement of settling and dewatering properties

The conditions during settling in secondary settling tanks can be best simulated by the measurement of zone settling velocity. The measurement should be performed in transparent cylinders so that the speed of movement of the sludge-supernatant interface can be read. The higher and wider the cylinder is, the more realistic figures we obtain. A typical settling curve and settling phases are shown in Figure 3.14. The zone settling velocity is calculated from the slope of the curve in the phase of zone settling (II).

In wastewater treatment practice, however, the measurement of the so-called sludge volume index (SVI) is the most common way of characterizing activated sludge settleability. The standard sludge volume index is defined as follows:

$$\mathrm{SVI} = V_{30}/X \tag{3.5}$$

where

SVI = sludge volume index (ml/g)
V_{30} = volume of settled sludge after 30 minutes sedimentation
X = concentration of activated sludge (mixed liquor suspended solids) (g/l).

The numerical value of SVI is affected by many factors, especially by sludge concentration or the volume of sludge after 30 min of sedimentation and by the so-called wall effect in the settlometer. Therefore in some countries the conditions of the settling test are standardized in a certain way. The common modifications are:

- stirred sludge volume index (SSVI); the measuring cylinder is equipped with a slowly rotating stirring device
- SVI at standard concentration; the test is performed at a defined sludge concentration, for instance 3.5 g/l ($SVI_{3.5}$)
- diluted sludge volume index (DSVI); the volume of settled sludge after 30 min of sedimentation should not exceed 200 ml l^{-1}; if it does, the test is repeated with diluted sludge.

Activated sludges can be classified according to the zone settling velocity and the sludge volume index, as shown in Table 3.2.

When the sludge separation problems are caused by filamentous microorganisms, it is necessary to quantify the filamentous microorganisms in activated sludge. For that purpose the measurement of **total extended filament length** (TEFL) was developed by Sezgin *et al.* (1978). The total length of filaments protruding from flocs or freely floating in the bulk liquid is measured. TEFL values of 10^7 m/ml or 10^4 m/g are considered a boundary between non-bulking and bulking activated sludges.

The measurement of TEFL is rather laborious and time-consuming. For a routine examination of activated sludges in wastewater treatment plants, a rapid and simple method of subjective scoring of filament abundance was developed by Jenkins *et al.* (1986). The abundance of filaments is classified according to the scale given in Table 3.3, from none to excessive.

When there is an increased abundance of foam-forming nocardioform actinomycetes, of rather irregular shape and frequently branched (see Figure 3.13d), Pitt and Jenkins (1990) recommend a modified method based on counting the number of intersections of nocardia filaments with a hairline in the eyepiece.

3.4.3.2 Activated sludge separation problems

Activated sludge should exhibit certain features that are of primary importance to the operation activated sludge system. Well-settling activated sludge:

- settles fast, with zone settling velocities of 1 m/h and more
- does not occupy an excessive volume after settling and thickening in secondary clarifier
- after sedimentation leaves a clear supernatant (X = 15 mg/l and less)
- does not rise within at least a 2–3 h period after sedimentation.

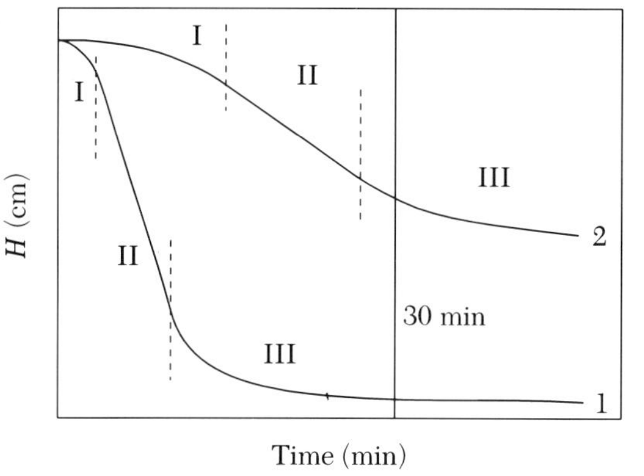

Figure 3.14. Example of activated sludge settling curves. Curves: 1, well settling sludge; 2, poorly settling sludge. Regions: I, reflocculation; II, zone settling; III, transition and compaction.

In wastewater treatment practice, six major problems related to microbial biomass quality can be distinguished (Jenkins *et al.* 1986; Wanner 1994).

(i) *Dispersed growth.* The activated sludge microorganisms are dispersed freely in the cultivation medium as individual cells or small clumps with a diameter of up 10–20 μm. The sedimentation rate of individual cells or bacterial clumps is too low for gravity sedimentation; no zone settling occurs in secondary settling tanks. This has two impacts on activated sludge process:

- the separation efficiency of the secondary settling tank is very low; the final effluent is turbid
- because of poor separation efficiency, a significant amount of biomass escapes from the system. Therefore only low values of sludge age (Θ_X) can be maintained in the system. The system with dispersed growth resembles a chemostat more than a continuous cultivation system with biomass recycle.

The poor bioflocculation is caused by a low production of extracellular biopolymers (glycocalyx) creating a matrix of firm activated sludge flocs. One of the typical reasons for the dispersed growth is a very high organic loading of biomass when bacteria are not forced to produce glycocalyx. Another reason may be toxicity of treated wastewater.

(ii) *Unsettleable microflocs*. The outer symptoms of this case of separation problems look very similar to dispersed growth at first sight:

Table 3.2. Types of activated sludge according to their settleability

Type of sludge	SVI (ml/g)	ZSV (m/h)
Well-settling	<100	>3
'Light'	100–200	2–3
Bulking	>200	<1.2

ZSV is defined as zone settling velocity.

Table 3.3. Subjective scoring of filament abundance (Jenkins et al. 1986)

Numerical value	Abundance	Description of microscopic picture
0	None	
1	Few	Filaments present, but only observed in an occasional floc
2	Some	Filaments commonly observed, but not present in all flocs
3	Common	Filaments observed in all flocs, but at low density (e.g. 1-5 filaments per floc)
4	Very common	Filaments observed in all flocs at medium density (e.g. 5-10 filaments per floc)
5	Abundant	Filaments observed in all flocs at high density (e.g. > 20 filaments per floc)
6	Excessive	Filaments present in all flocs: appears more filaments than flocs and/or filaments growing in high abundance in bulk solution

the final effluent from secondary clarifier is not clear and contains many microparticles of escaping biomass. However, the nature of the problem is different, as can be seen under the microscope. The unsettleable particles are of larger dimension (*ca.* 50–100 μm) than in dispersed growth; the particles are roughly spherical and compact. These microflocs result from the disintegration of initially firm and sound flocs.

During the settling test the activated sludge is quickly divided into two parts. The larger flocs settle rapidly and when the sludge volume index is calculated on the basis of the volume of these larger flocs, its value is quite low (around 50 ml/g). However, the supernatant in the cylinder is turbid and a substantial fraction of total biomass remains in these unsettleable particles.

The reasons for the disintegration of activated sludge flocs are:

- insufficient production of glycocalyx, or its consumption by bacteria inside the flocs owing to a low organic loading of biomass (typical of high sludge age systems, extended aeration)
- absolute absence of filamentous microorganisms that according to Jenkins *et al.* (1986) form a 'backbone' of most flocs larger than 80–100 μm
- disintegration by shearing effects, for instance inappropriate mechanical aerators.

(iii) *Rising sludge*. When this problem occurs, the water surface in the secondary clarifier is covered with patches (or in worse cases completely) of floating activated sludge. When the phenomenon is observed in a glass cylinder, two phases can be distinguished:

- first the activated sludge settles rapidly and a rather compact bottom layer of settled sludge and a clear supernatant are formed
- after a certain period of time (at elevated temperatures, even less than 30 min, which can cause difficulties in the SVI test), a part or the whole volume of the settled and thickened sludge starts to float and move up to the water surface.

The floating material is full of gas bubbles. The nature of this phenomenon is endogenous denitrification that takes place in the settled and thickened layer of activated sludge. Because of a high biomass concentration, the dissolved oxygen from the previous aeration is quickly depleted and anoxic conditions can thus be established provided that nitrification occurs in the system. The bubbles of nitrogen liberated during this endogenous denitrification act as sludge 'carrier'.

(iv) *Viscous bulking*. This rather broad term describes the consequences but not the causes. The activated sludge contains an excessive amount of extracellular biopolymers, which impart a slimy, jelly-like consistency to the sludge. As the biopolymers are hydrophilic colloids, the activated sludge becomes highly water retentive. Such a 'hydrous' activated sludge exhibits low settling and compaction velocities. The biopolymers are also natural surface active agents. When the viscous activated sludge is intensively aerated, a strong foaming can appear.

The production of biopolymers is characteristic of most floc-forming microorganisms but under normal conditions (no toxic compounds, nutrient balanced growth) the quantity of biopolymers generated is just enough for the formation of firm flocs. In contrast, zoogloeal bacteria produce always large quantities of biopolymers because the individual cells of *Zoogloea* are fixed in slimy colonies. Thus some authors use the term 'zoogloeal bulking' for describing the settling problems caused by an excessive presence of highly hydrated zoogloeal colonies in activated sludge.

(v) *Filamentous bulking*. This is a typical problem of poor compaction of activated sludge, which results in:

- low return and waste activated sludge concentrations
- difficulties in maintaining the required activated sludge concentration in reaction basins
- poor sludge dewaterability
- hydraulic overloading of sludge-handling facilities.

Filamentous microorganisms interfere with the sedimentation and compaction of activated sludge flocs in two ways (Jenkins *et al.* 1986).

(1) Some kinds of filamentous microorganism prefer to grow in the interiors of flocs, and produce flocs with a very diffuse open structure. These open flocs provide a lot of space for water inside them, so that although the aggregation of individual flocs is not mechanically hindered by filaments protruding from the flocs, too much water remains 'captured' in the settled sludge.

(2) The second way in which filaments can deteriorate the sedimentation and compaction of activated sludge flocs is much more common. Most filamentous microorganisms observed in activated sludges protrude from rather compact and firm flocs into the bulk liquid. The filaments, which in low numbers form a backbone of firm flocs, in large number are able to prevent mechanically the compaction of individual flocs.

(vi) *Foaming caused by filamentous microorganisms*. The biological foaming by 'foam-forming' filamentous microorganisms is a complex of physico-chemical and biochemical processes leading to the stabilization of a three-phase system: air–water–microbial cells. The stabilization of biological foams results from the following features of foam-forming filaments:

- production of extracellular materials such as lipids, lipopeptides, proteins and carbohydrates that have the properties of surface active agents (biosurfactants)
- in contrast with other filaments and floc-formers, the cell walls of foam-forming microorganism are strongly hydrophobic.

The formation of stable foams in the aeration basins of activated sludge plants can create a wide range of operation problems:

- aesthetic problems, slippery path along the aeration basins covered by escaping foam
- floating biomass in secondary clarifier, deteriorating the final effluent quality
- accumulation of significant amounts of biomass into the foam, which is not available for treatment processes; loss of possibility of controlling activated sludge age.

3.5 Selection mechanisms in activated sludge

The actual microbial composition of activated sludge is a result of strong competition for energy and carbon sources, nutrients and sometimes also electron acceptors (e.g. so-called low dissolved oxygen filamentous bulking; see Jenkins *et al.* (1986)). The competition is influenced by numerous factors, both intrinsic to the activated sludge process, such as *biomass retention time*, *actual substrate*, *oxygen and nutrients concentrations in reactor* or *cultivation conditions* (oxic, anoxic, anaerobic) and factors from outside the activated sludge system. These external factors such as *wastewater composition*, *temperature* and *pH*, are difficult to control but their effect must be considered in control strategies for wastewater treatment plants. The following basic mechanisms are decisive for the selection of microbial species in activated sludge (Wanner 1994).

3.5.1 Kinetic selection

This selective pressure results from differences in growth and substrate utilization rates. In **balanced growth** (readily biodegradable substrate uptake and cell growth occur simultaneously) two types of competition strategy are possible (in terms of saturation kinetics):

(1) *r*-strategy: maximum specific growth and substrate utilization rates at high substrate concentrations.

(2) *K*-strategy: high substrate affinity at low substrate concentrations.

Floc-forming microorganisms in activated sludge are generally the *r*-strategists, whereas most filamentous microorganisms in activated sludge belong to the *K*-strategists. However, so-called **unbalanced growth**, when the phases of uptake and utilization of substrate (both soluble and particulate) or nutrients are partly or fully separated, is a more common case in the activated sludge process. In competition under conditions of unbalanced growth, the microorganisms that can sequester substrate more rapidly from the bulk liquid will be selected provided that there is time enough for the generation of **accumulation/storage** capacity in the cells of these microbes. In activated sludge the accumulation/storage and regeneration selection mechanism is typical of most floc-forming microorganisms. The process of substrate accumulation/storage takes place in the contact zone of the activated sludge system (see below), which can be not only oxic but also anoxic or anaerobic, whereas endogenous metabolism (regeneration) occurs in the main aeration basin or in a separate regeneration zone placed in the return sludge stream.

The results of competition of floc-forming and filamentous microorganisms on the basis of kinetic selection are influenced by the following factors.

3.5.1.1 Wastewater composition

Wastewater is a source of substrates, nutrients and micronutrients for the microbial consortium of activated sludge. In addition, wastewater continuously inoculates this consortium with bacteria growing in sewers.

Readily biodegradable substrates. Low molecular mass organic compounds, i.e. compounds with simple molecules, can usually be directly utilized by bacterial cells. These compounds (e.g. monosaccharides, alcohols, volatile fatty acids and amino acids) represent approximately 10–20% of the COD of common municipal wastewaters. Therefore their concentration in the mixed liquor of activated sludge is usually not very high, which is a favourable factor for the growth of some filamentous microorganisms (*K*-strategists). In particular, Types 021N, *Sphaerotilus natans*/1701, 0041 and *H. hydrossis* can be expected in systems treating wastewaters with elevated concentrations of these compounds. Special attention has to be paid to the design of activated sludge systems for wastewaters with an increased fraction of industrial wastewaters (food industry). Reduced sulphur compounds can also be easily metabolized and thus support the growth of sulphur-metabolizing filamentous organisms such as Type 021N/*Thiothrix* or *Beggiatoa*.

Particulate, slowly biodegradable substrates. Most of the utilizable organic compounds in municipal wastewaters can be described as the so-called particulate substrate. The particulate substrate is formed by large-molecule organic compounds, which are present in the wastewater either as colloids or as true suspended solids. In both cases those molecules have to be attacked by extracellular enzymes before they become available to cells. The mechanism of particulate substrate disintegration is generally termed **hydrolysis** and the products of this process are very similar, from a chemical point of view, to readily biodegradable substrates originally present in wastewaters. However, hydrolysis is a surface phenomenon and is connected with flocs. Therefore, if the main aeration basins, where the hydrolysis occurs, are not operated as **completely mixed tanks**, the readily biodegradable substrates released by hydrolysis will be more available to floc-formers than to filamentous microorganisms (Ekama and Marais 1986; Gujer and Kappeler 1992). However, some special kinds of particulate substrates can support the growth of filamentous microorganisms. This is especially true of fats and grease, which are selectively concentrated in foams and thus support the growth of nocardioform actinomycetes. Long-chain fatty acids are reported in the literature as a specific substrate supporting the growth of *Microthrix parvicella* (Slijkhuis and Dienema 1982). Oleic compounds in wastewater can also support the so-called viscous bulking caused by zoogloeal bacteria, especially in activated sludge systems with a high substrate concentration gradient.

Inoculation of activated sludge systems from wastewaters. The sewerage network and wastewater treatment plants form one system and the processes in sewers can significantly affect the composition of activated sludge consortia. In particular, in extended sewer systems readily biodegradable substrates can be taken up by bacteria growing in inner slimes. These wall growths are continuously sloughed off and inoculate the treatment plant. In most cases the sewer biofilms consist of non-filamentous bacteria but filaments such as *Sphaerotilus natans* can be a significant component of that inoculum (Kappeler and Gujer 1992). Although nothing can be done to prevent the growth of filaments in sewers, it is always useful to know whether the activated sludge system is inoculated with floc-formers or with filaments.

3.5.1.2 Biomass retention time

Biomass retention time, Θ_X ('sludge age'), has a great impact on the distribution of individual microbial species in the consortium of activated sludge according to their growth and decay rates. A small value for Θ_X might result in a wash-out of slow-growing species whose net growth rate $(m - b)$ is slower than the D, the dilution rate $(D = 1/\Theta_X)$. In contrast, high values for Θ_X favour the slow-growers in the mixed culture of activated sludge. Therefore the designer should realize that at different regions of Θ_X different filamentous microorganisms will predominate in activated sludge. Unfortunately, the 'safe' values of Θ_X, when all filamentous microorganisms are certainly washed out, are too low to be applied as a common bulking control measure. The distribution of the most common filamentous microorganisms according to biomass retention time is given in Figure 3.15. This distribution is strongly affected by temperature: at elevated temperatures the 'wash-out' biomass retention time is lower. The last group of filaments preferably growing at high Θ_X values is also termed 'low F/M' filaments (Jenkins *et al.* 1986). In IAWQ nomenclature the term 'F/M' is equivalent to specific activated sludge loading B_X (although some authors use the F/M ratio in the sense of S_0/X_0 ratio). However, at present the 'long sludge age' (or 'low F/M') activated sludge systems are operated mostly as nutrient removal systems, so that it is difficult to attribute the appearance of 'low F/M' filaments only to long biomass retention time because the

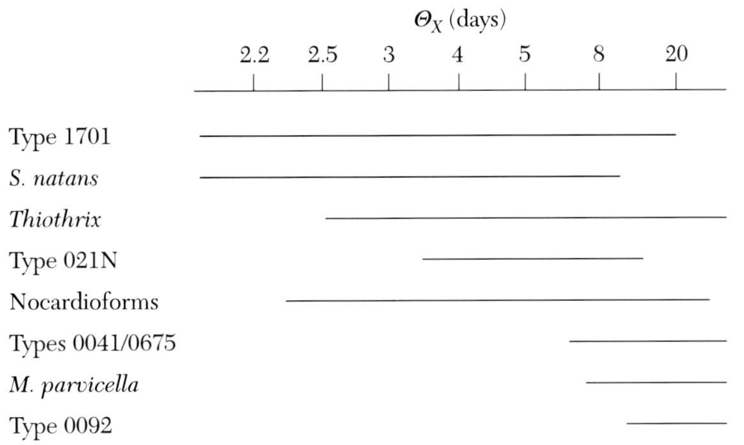

Figure 3.15. Distribution of filamentous microorganisms according to Θ_X (adapted from Jenkins et al. (1993) and Richard (1989)).

different cultivation conditions (anaerobic, anoxic) in the nutrient removal systems are also important in the selection of microorganisms in activated sludges.

3.5.1.3 Actual substrate concentration in reactor

The activated sludge cultured in reactors with substrate concentration gradients gains certain features that were described by Gabb *et al.* (1988) as a 'selector' effect:

(1) high rates of substrate consumption
(2) high oxygen (or, generally, electron acceptor) uptake rate
(3) enhanced growth of zoogloeal bacteria.

When the activated sludge flocs are alternately exposed to higher and lower substrate concentrations, a significant stratification of the flocs can appear, which increases the diversity of metabolic processes occurring inside the flocs. Thus even in fully aerated activated sludge systems a significant fraction of available substrate can be utilized under anaerobic or anoxic conditions (Albertson 1987). The substrate concentration gradient then leads to a combined effect of kinetic and **metabolic selection**. Because of the floc stratification, a fourth feature can be added, namely:

(4) increased metabolic diversity.

3.5.1.4 Dissolved oxygen, nutrients, pH and temperature in aeration basins

Dissolved oxygen. Some filamentous microorganisms (e.g. *Sphaerotilus natans*/Type 1701, *H. hydrossis* according to Jenkins *et al.* (1986)) exhibit great affinity for dissolved oxygen at low concentrations because of low values of half-saturation constant K_0. The boundary between 'bulking' and 'non-bulking' DO concentrations is not fixed because this value depends on the actual value of activated sludge loading, B_X, in the reactor. For the estimate of DO concentrations required for avoiding low DO bulking, the data obtained by Palm *et al.* (1980) are summarized in Table 3.4. As can be seen from this table, at higher sludge loadings the 'safe DO concentrations are quite high. This is important fact for the design of aeration systems. Especially in the head-end of aeration basins the oxygenation capacity of the aerators must be high enough to guarantee the DO concentrations given in Table 3.4. The aeration equipment in oxic zones of nutrient-removal activated sludge systems should convert the mixed liquor from preceding anaerobic or anoxic conditions to oxic conditions as quickly as possible.

Nutrients. As with dissolved oxygen, some filaments also exhibit a higher affinity for nutrients (i.e. nitrogen, phosphorus and micronutrients). However, the experiments with Type 021N performed by Richard *et al.* (1985) showed that the dominance of filamentous microorganisms under low nutrient concentrations is not only a matter of the *K*-strategy but that the filaments might exhibit a kind of 'accumulation' capacity for nutrients, which gives them another advantage under conditions of unbalanced growth. Therefore the addition of nutrients for controlling filamentous bulking should be sufficient to saturate this accumulation capacity. The exact doses of limiting nutrient(s) can be assessed only by pilot tests. When the tests cannot be performed, Richard (1989) recommends the maintenance of these effluent concentrations: total inorganic *nitrogen* at least *1 mg/l*, soluble orthophosphate *phosphorus* more than *0.2 mg/l*.

The possibility of filamentous bulking by nutrient limitation shows the importance of the requirement to know the composition of the treated wastewater before starting the design of an activated sludge system, especially if there is an important contribution by industrial wastewaters. A severe lack of nutrients may also result in serious problems with *viscous* bulking because an elevated synthesis of extracellular

Table 3.4. Relationship between BX and 'safe' DO concentrations in oxic contact zones

B_X (kg kg^{-1} day^{-1})	DO (mg/l)
0.3	1.0
0.5	2.0
0.75	3.0
0.9	4.0

polymers is one of the reactions of microbial cells to such unfavourable conditions.

pH. None of the most common filamentous microorganisms exhibits a special preference for extreme pH values. Some micromycetes prefer low pH values; however, such a decrease in pH in municipal wastewater treatment plants (e.g. by nitrification) leading to excessive fungal growth is improbable. Lime-dosing equipment should be considered when some acid stream might be discharged to the plant.

Temperature. Temperature significantly affects the rates of all biochemical processes as well as the solubility of oxygen in the mixed liquor. Thus elevated temperatures support the growth of filamentous microorganisms connected with low DO concentrations. The oxygenation capacity of aeration systems should be calculated for the highest mixed liquor temperatures expected during the year to guarantee the DO concentrations from Table 3.4 in warm weather also.

There are significant seasonal shifts in the dominance of individual filamentous types. Typically, *Microthrix parvicella*, which dominates the filamentous population in winters, is replaced by nocardioform actinomycetes, Type 0041 or *Nostocoida limicola* in warm seasons. This shift is certainly connected with temperature changes but an exact explanation of this phenomenon is still not available. According to European experience, more severe *Microthrix parvicella* foaming problems are to be expected in winter than in summer (Seviour *et al.* 1990; Kristensen *et al.* 1991; Foot 1992).

3.5.2 Metabolic selection

In nutrient-removal activated sludge systems the decisive fraction of readily biodegradable substrates from wastewater is utilized not under oxic but in anaerobic and/or anoxic conditions. Thus the microorganisms able to metabolize substrates under these conditions will be selected. This is the principle of **metabolic selection**.

The alternation of anaerobic and oxic cultivation conditions is a prerequisite for the enhanced biological phosphorus removal (EBPR) mechanism in nutrient removal systems (see Figure 3.7). As the filamentous microorganisms from groups S and C (see above) are unable to utilize substrate under alternating anaerobic/oxic cultivation conditions with a rate comparable to that of floc-forming microbes, their suppression in nutrient-removal activated sludge systems is based on metabolic principles. Whereas the group S and C filamentous microorganisms were quite common in conventional activated sludge plants, recent surveys of filaments in nutrient removal systems done in Europe, South Africa, the U.S.A. and Australia indicate that these filaments are not important in systems with anaerobic and/or anoxic zones. The elimination of group S and C filaments in systems with anoxic zones is based on different denitrification rates between filaments and floc-formers.

Although studied in detail, the abundant occurrence of group A filaments in nutrient removal systems has not been satisfactorily explained yet.

3.5.3 Selective predation

Protozoa and metazoa can significantly affect the results of competition between bacterial species. For instance, Verhagen and Laanbroek (1992) observed a selective predation by the flagellate *Adriomonas peritocrescens* on nitrifying bacteria *Nitrosomonas europea* and *Nitrobacter winogradskyi*. Cech *et al.* (1991, 1994) found that the competition between poly-P bacteria and the so-called 'G' bacteria was seriously affected by protozoan and metazoan grazing. The 'G' bacteria forming large compact colonies were predation-resistant and outcompeted the predation-susceptible poly-P bacteria.

Recent experiments in Japan (Inamori *et al.* 1991) indicated that there are some 'spaghetti eaters' among predatory ciliates, which specialize in consuming filamentous microorganisms.

3.6 Application of population dynamics principles to the design of nutrient-removal activated sludge plants

3.6.1 Plant configuration

The design of activated sludge system has to combine all factors favouring the growth of floc-formers:

- substrate concentration gradient in the system (or at least in its head-end)
- rapid accumulation ('biosorption') of substrate with subsequent regeneration of accumulation/storage capacity
- in nutrient removal systems, the utilization of most substrate under anaerobic and/or anoxic conditions.

There are four principal process configurations effectively supporting the growth of floc-formers. In these configurations the basic principles of kinetic selection are employed either

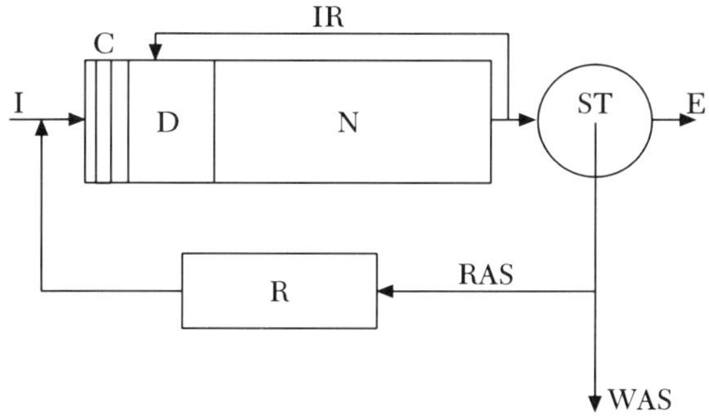

Figure 3.16. Schematic of the so-called R-D-N Process: C, compartmentalized anoxic contact zone; D, denitrification zone; N, nitrification zone; ST, settling tank; R, regeneration zone; I, influent; E, effluent; RAS, return activated sludge; WAS, waste activated sludge; IR, internal recirculation.

alone or in combination with metabolic selection:

(1) sequencing batch reactor with short filling period ('dump-fill') or with an inlet mixing zone
(2) continuous plug-flow reactors; as even the long corridor-type reactors exhibit a high degree of longitudinal mixing, the reactor should be compartmentalized to approach the plug-flow hydraulic regime
(3) completely mixed or compartmentalized continuous-flow reactor with a contact zone placed ahead of the main reactor; the contact zone should preferably be divided into two to four compartments
(4) activated sludge system with return sludge regeneration; the main reactor should be arranged in configurations 2 or 3.

In all the above configurations, the return activated sludge can be mixed with the treated wastewater under either oxic or anaerobic/ anoxic conditions. An example of activated sludge configuration combining most of the factors favouring the growth of floc-formers is given in Figure 3.16.

3.6.2 Design parameters

3.6.2.1 Oxic contact zone design

The design criteria for the oxic contact zone can also be applied to the design of the inlet part of compartmentalized plug-flow systems or to the design of SBRs. However, the oxic contact zone (originally patented as the so-called selector in Czechoslovakia by Chudoba *et al.* (1972)) can be used only in conventional activated sludge systems. This concept cannot be employed in biological nutrient removal (BNR) plants because the readily biodegradable substrate necessary for the EBPR and denitrification is 'inefficiently' consumed under oxic conditions. The following parameters should be considered in the design of oxic contact zones:

Table 3.5. Recommended contact time in oxic contact zones

Source	Θ_{CT} (minutes)
van Niekerk *et al.* (1987)	12–18
ATV Report 2.6.1 (1989)	10–12
Daigger *et al.* (1985)	15
Eikelboom (1991b)	10–15
Pujol (1992)	15
Rensink and Donker (1993)	≥10

Contact time. The contact time is based on the flow of mixed liquor and not only on the wastewater flow. The parameter is directly connected with the size of contact zone. When the contact time is insufficient, soluble substrate is not consumed in the contact zone and penetrates to the main reactor where it can support the growth of filaments. In contrast, when the contact time is too long, the growth of filaments occurs right in the contact zone, especially when this is not compartmentalized. The recommended values of contact time are summarized in Table 3.5.

The recommended values of contact time lie in a range of 10–20 min. For conventional activated sludge systems this contact time results in the volume ratio V_{CZ}/V_{total} of about 1/10. This ratio is also sufficient for a full restoration of accumulation/storage capacity in the main reactor. The recirculation ratio of return activated sludge entering the contact zone should be less than 1.

Activated sludge loading in the contact zone. The activated sludge loading B_X was selected as a design criterion instead of 'floc loading'. Although this parameter describes more precisely the relationship between actual substrate and biomass mass fluxes, for continuous-flow systems the activated sludge loading is more convenient for calculations. Table 3.6 brings the figures from the most successful case histories of bulking control found in the literature. The term 'aerated' contact zone means that as well as oxic conditions in the bulk liquid, anaerobic and anoxic conditions are possible in the flocs under such high loadings. All B_X values are related to the *whole* volume of the contact zone, so that for compartmentalized contact zones the actual sludge loading in the first compartment is correspondingly higher. The asterisked values were calculated for MLVSS (organic fraction of X).

Conditions in the main reactor. The selection of floc-formers does not occur only in the contact zone; the conditions in the main reactor after the contact zone are part of the selection process (in this sense the term 'selector' should cover the whole configuration and not only the contact zone). The retention time of activated sludge in the main aeration basin is sufficient

Table 3.6. Activated sludge loading in aerated contact zones

Source	B_X (kg kg^{-1} day^{-1}), based on	
	BOD$_5$	COD
Albertson (1991)	3.0	
ATV Report 2.6.1 (1989)	3.0	20°
Chudoba and Wanner (1989)	>3.0	
Daigger and Nicholson (1990)	3.2–4.9	
Eikelboom (1991b)	2.0–5.0	
Lee *et al.* (1982)		>20°
Linne *et al.* (1989)	5.0–6.0	

° Limit by which the contact zone COD should not exceed effluent COD.

for a complete restoration of accumulation/storage capacity when the activated sludge loading in the whole system does not exceed 0.3–0.5 kg kg^{-1} d^{-1} (BOD$_5$, X) (ATV Report 2.6.1 1989; Chudoba and Wanner 1989; Eikelboom 1991). If the main reactor is a tank with high dispersion number, the soluble COD in the effluent from the contact zone should not exceed the final effluent soluble COD by more than 20–30 mg/l.

3.6.2.2 Design of anaerobic and anoxic contact zones of BNR systems

The primary task of anaerobic and anoxic contact zones in BNR activated sludge systems is to create conditions for the elimination of nitrogen and phosphorus. Thus the design of these zones has to follow principally the guidelines for designing BNR systems (see, for example, Randall *et al.* 1992). However, the designer should always try to implement bulking control strategies in the design, although the conditions for bulking control can sometimes be contradictory to the requirements of BNR processes.

Anaerobic zone. The anaerobic contact time sufficient for bulking control is about 0.5–1.0 h, which could be too short for the EBPR. The value of activated sludge loading and the existence of a substrate concentration gradient in anaerobic contact zones do not play as important a role in bulking control as do those in oxic and anoxic contact zones. The floc-formers growing in systems with anaerobic zones differ from those in systems with oxic and anoxic zones (Jenkins 1992a) and the competition between these floc-formers and filaments is based mostly on metabolic selection. The B_X of 1 kg kg^{-1} d^{-1} (BOD$_5$, X) in the anaerobic zone was enough for an efficient bulking control in a brewery wastewater treatment plant (Wanner and Roskota 1993).

Anoxic zone. A certain analogy with the design of oxic contact zones can be used because the floc-formers supported by anoxic contact zones are mostly the same microbes that are favoured by oxic contact zones. From the successful case histories described in the available literature, the Southerly Wastewater Treatment Plant in Columbus, Ohio, can be mentioned as the most typical design example (Albertson 1991). The reactor consists of ten compartments, of which the first three are anoxic. The activated sludge loading in the first compartment is B_X = 6 kg kg^{-1} d^{-1} (BOD$_5$, X).

To avoid a 'secondary' growth of filamentous microorganisms (Oxic Zone Growers) in the oxic part of the BNR system, Jenkins (1992b) required the soluble COD in the effluent from anoxic zone to be less than 60 mg/l (for municipal wastewaters this requirement corresponds well with the above recommended difference between soluble COD values in the effluent from contact zone and in the final effluent). The sizing of the anoxic contact zone can therefore be based on the COD value necessary for achieving the effluent soluble COD of 60 mg/l. The following typical figures are recommended for the calculation:

- specific COD consumption for nitrate nitrogen reduction 8 g/g (COD, NO$_3$-N)
- specific denitrification rate $r_{X,den}$ = 3 mg g^{-1} h^{-1} (NO$_3$-N, X) or 4 mg g^{-1} h^{-1} (NO$_3$-N, X$_{org}$).

The volume of anoxic contact zone obtained with the above figures results in contact times between 10 and 30 min. This value of anoxic contact time is considered to be a good design practice in the U.S.A. (Hsu and Wilson 1992).

3.7 Conclusions

Microbial population dynamics studies the phenomena that govern the relationships between individual groups of microorganisms in complex consortia used for biological wastewater treatment and between the microorganisms, and external factors such as wastewater composition, cultivation conditions and temperature. The basic biochemistry and microbiology of the following main processes were described:

- oxic organic carbon removal
- anoxic carbon removal – denitrification
- nitrification
- enhanced biological phosphorus removal
- conversion of sulphur compounds.

From the point of view of good activated sludge separation properties, the competition between filamentous and floc-forming microorganisms seems to be the most important component of activated sludge population dynamics. There are two major selective mechanism in activated sludge, namely:

- kinetic selection based on different rates of substrate utilization

- metabolic selection resulting from different abilities of microorganisms to use substrate under anoxic and anaerobic cultivation conditions.

The configuration of a contemporary nutrient-removal activated sludge system must combine all factors favouring the growth of floc-forming microorganisms:

- substrate concentration gradient in the system (or at least in its head-end)
- rapid accumulation ('biosorption') of substrate with subsequent regeneration of accumulation/storage capacity
- utilization of most substrate under anaerobic and/or anoxic conditions.

The design criteria based on successful case histories have been summarized in this chapter.

3.8 References

Albertson, O.E. (1987) The control of bulking sludges: from the early innovators to current practice. *J. Wat. Pollut. Control F.* **59** (4), 172–182.

Albertson, O.E. (1991) Bulking sludge control – progress, practice and problems. *Wat. Sci. Technol.* **23** (4/5), 835–846.

Anon. (1969) Milwaukee mystery: unusual operating problems develop. *Wat. Sew. Wks* **119** (6), 213.

ATV Working Group 2.6.1 Report (1989) Prevention and control of bulking sludge and scum. *Korresp. Abwass.* **36** (13), 165–177.

Blackall, L.L. (1994) Molecular identification of activated sludge foaming bacteria. *Wat. Sci. Technol.* **29** (7), 35–42.

Blackbeard, J.R., Gabb, D.M.D., Ekama, G.A. and Marais, G.v.R. (1988) Identification of filamentous organisms in nutrient removal activated sludge plants in South Africa. *Wat. SA* **14** (1), 29–33.

Bock, E., Wilderer, P.A. and Freitag, A. (1988) Growth of *Nitrobacter* in the absence of dissolved oxygen. *Wat. Res.* **22** (2), 245–250.

Brodisch, K.E.U. (1985) Interaction of different groups of micro-organisms in biological phosphate removal. *Wat. Sci. Technol.* **17**, 89–97.

Brodisch, K.E.U. and Joyner, S.J. (1983) The role of micro-organisms other than *Acinetobacter* in biological phosphate removal in activated sludge processes. *Wat. Sci. Technol.* **15**, 117–125.

Buchan, L. (1984) Microbiological aspects. In *Theory, design and operation of nutrient removal activated sludge processes* (ed. H.N.S. Wiechers *et al.*), pp.9/1–9/6. Pretoria, SA: Water Research Commission.

Casey, T.G., Wentzel, M.C., Ekama, G.A., Loewenthal, R.E. and Marais, G.v.R. (1994) An hypothesis for the causes and control of anoxic–aerobic (AA) filament bulking in nutrient removal activated sludge systems. *Wat. Sci. Technol.* **29** (7), 203–212.

Cech, J.S., Hartman, P., Macek, M. (1994) Bacteria and protozoa population dynamics in biological phosphate removal systems. *Wat. Sci. Technol.* **29** (7), 109–117.

Cech, J.S., Macek, M. and Wanner, J. (1991) Limits of enhanced biological phosphate removal process. In *Biological Approach to Sewage Treatment Process: Current Status and Perspectives* (ed. P. Madoni), pp. 437–439. Luigi Bazzuchi Center, Perugia, Italy.

Chiesa, S.C. and Irvine, R.L. (1985) Growth and control of filamentous microbes in activated sludge: an integrated hypothesis. *Water Res.* **19** (4), 471–479.

Chudoba, J. and Wanner, J. (1989) Notes on oxic, anoxic, and anaerobic selectors. *Newsletter of the IAWPRC Specialist Group on Activated Sludge Population Dynamics* **1** (2), 46–49.

Chudoba, J., Grau, P. and Dohanyos, M. (1972) Czechoslovak patent 167,570.

Cloete, T.E. and Steyn, P.L. (1988) A combined membrane filter–immunofluorescent technique for the *in-situ* identification and enumeration of *Acinetobacter* in activated sludge. *Wat. Res.* **22** (8), 961–969.

Daigger, G.T. and Nicholson, G.A. (1990) Performance of four full-scale nitrifying wastewater treatment plants incorporating selectors. *Res. J. Wat. Pollut. Control Fed.* **62** (5), 676–683.

Daigger, G.T., Robbins, M.H. and Marshall, B.R. (1985) The design of a selector to control low F/M filamentous bulking. *J. Wat. Pollut. Control Fed.* **57** (3), 220–226.

Eikelboom, D.H. (1975) Filamentous organisms observed in activated sludge. *Wat. Res.* **9**, 365–388.

Eikelboom, D.H. (1982) Biosorption and prevention of bulking sludge by means of a high floc loading. In *Bulking of activated sludge: preventative and remedial methods* (ed. B. Chambers and E.J. Tomlinson), pp. 224–242. Ellis Horwood Limited, Chichester.

Eikelboom, D.H. (1991a) The role of competition between flocforming and filamentous bacteria in bulking activated sludge. In *Biological approach to sewage treatment process: current status and perspectives* (ed. P. Madoni), pp. 143–149. Luigi Bazzucchi Center, Perugia.

Eikelboom, D.H. (1991b) Scum formation in carrousel treatment plants. In *Proceedings of the IAWPRC Specialised Seminar on Interactions of Wastewater, Biomass and Reactor Configurations in Biological Treatment Plants*, 21–23 August 1991, Copenhagen, paper no. 24.

Eikelboom, D.H. and van Buijsen, H.J.J. (1981) *Microscopic Sludge Investigation Manual*. IMG–TNO Report A94a, Delft.

Ekama, G.A. and Marais, G.v.R. (1986) The implication of the iawprc hydrolysis hypothesis on low F/M bulking. *Wat. Sci. Technol.* **18**, 11–19.

Ekama, G.A., van Haandel, A.C. and Marais, G.v.R. (1979) The present status of research on nitrogen removal. Report no. W29, University of Cape Town.

Foot, R.J. (1992) Effects of process control parameters on the composition and stability of activated sludge. *J. Inst. Wat. Envir. Mgmt.* **6** (2), 215–228.

Gabb, D.M.D, Still, D.A., Ekama, G.A., Jenkins, D., Wentzel, M.C. and Marais, G.v.R. (1988) Development and full scale evaluation of preventative and remedial methods for control of activated sludge bulking. Research Report W62, University of Cape Town.

Grabinska-Loniewska, A. (1991) Denitrification unit biocenosis. *Water Res.* **25** (12), 1565–1573.

Gujer, W. and Kappeler, J. (1992) Modelling population dynamics in activated sludge systems. *Wat. Sci. Technol.* **25** (6), 93–103.

Hart, M.A. and Melmed, L.N. (1982) Microbiology of nutrient removing activated sludge. *Wat. Sci. Technol.* **14** (9/11), 1501–1502.

Hsu, M. and Wilson, T.E. (1992) Activated sludge treatment of municipal wastewater – U.S.A. practice. In *Activated sludge process design: theory and practice* (ed. W.W. Eckenfelder and P. Grau), pp. 37–68. Technomic Publishing Co., Lancaster, PA.

Inamori, Y., Kuniyasu, Y., Sudo, R. and Koga, M. (1991) Control of the growth of filamentous microorganisms using predacious ciliated protozoa. *Wat. Sci. Technol.* **23** (4/6), 963–971.

Jenkins, D. (1992a) Towards a comprehensive model of activated sludge bulking and foaming. *Wat. Sci. Technol.* **25** (6), 215–230.

Jenkins, D. (1992b) Bulking causato da bassi valori del carico organico. In *Il controllo del bulking negli impianti a fanghi attivi* (ed. L. Cingolani and R. Ramadori), pp. 43–54. Centro Luigi Bazzucchi, Perugia.

Jenkins, D., Richard, M.G. and Daigger, G.T. (1986) *Manual on the causes and control of activated sludge bulking and foaming*. Water Research Commission, Pretoria.

Jenkins, D., Richard, M.G. and Daigger, G.T. (1993) *Manual on the causes and control of activated sludge bulking and foaming*, 2nd edn. Lewis Publishers, Boca Raton.

Kappeler, J. and Gujer, W. (1992) Bulking in activated sludge systems: a qualitative simulation model for *Sphaerotilus natans*, Type 021N and Type 0961. *Wat. Sci. Technol.* **26** (3/4), 473–482.

Kavanaugh, R.G. and Randall, C.W. (1994) Bacterial populations in a biological nutrient removal plant. *Wat. Sci. Technol.* **29** (7), 25–34.

Kristensen, G.H., Jorgensen, P.E. and Henze, M. (1992) Characterization of functional microorganism groups and substrate in activated sludge and wastewater by AUR, NUR and OUR. *Wat. Sci. Technol.* **25** (6), 43–57.

Kristensen, G.H., Jorgensen, P.E. and Nielsen, P.H. (1991) Bundfaeldningsegenskaber for aktiv slam – undersogelse af danske renseanlaeg med naeringssaltfjernelse. Report of Water Quality Institute, Copenhagen and Aalborg University Centre, Laboratory for Environment, November 1991.

Kuba, T., Wachtmeister A., van Loosdrecht, M.C.M. and Heijnen J.J. (1994) Effect of nitrate on phosphorus release in biological phosphorus removal systems. *Wat. Sci. Technol.* **30** (6), 263–269.

Lee, S.-E., Koopman, B.L., Jenkins, D. and Lewis, R.F. (1982) The effect of aeration basin configuration on activated sludge bulking at low organic loading. *Wat. Sci. Technol.* **14**, 407–427.

Lindrea, K.C., Pigdon, S.P., Boyd, B. and Lockwood, G.A. (1993) Biomass characterization in a NDBEPR plant during start-up and subsequent periods of good and poor phosphorus removal. *Wat. Sci. Technol.* **29** (7), 91–100.

Linne, S.R., Chiesa, S.C., Rieth, M.G. and Polta, R.C. (1989) The impact of selector operation on activated sludge settleability and nitrification: pilot-scale results. *J. Wat. Pollut. Control Fed.* **61** (1), 66–72.

Lötter, L.H. and Murphy, M. (1985) The identification of heterotrophic bacteria in an activated sludge plant with particular reference to polyphosphate accumulation. *Wat. SA* **11** (4), 179–184.

McCarty, P.L. (1964) Thermodynamics of biological synthesis and growth. in *Proceedings of the 2nd International Conference on Advances in Water Pollution Research*, Tokyo, pp.169–199. Pergamon Press, Oxford.

McLeod, W.T. (ed.) (1984) *The New Collins Concise Dictionary of the English Language*, p. 348. Collins, London.

Painter, H.A. (1970) A review of literature on inorganic nitrogen metabolism in microorganisms. *Wat. Res.* **4** (5), 393–450.

Palm, J.C., Jenkins D. and Parker, D.S. (1980) Relationship between organic loading, dissolved oxygen concentration and sludge settleability in the completely-mixed activated sludge process. *J. Wat. Pollut. Control Fed.* **52** (10), 2484–2506.

Pitt, P. and Jenkins, D. (1990) Causes and control of *Nocardia* in activated sludge. *Res. J. Wat. Pollut. Control Fed.* **62** (2), 143–150.

Pujol, R. (1992) Control of filamentous bacteria by the contact zone: design and results. *Newsletter of the IAWQ Specialist Group on Activated Sludge Population Dynamics* **4** (1), 24–26.

Randall, C.W., Barnard, J.L. and Stensel, H.D. (eds) (1992) *Design and retrofit of wastewater treatment plants for biological nutrient removal.* Technomic Publishing Co., Lancaster, PA.

Rensink, J.H. and Donker, H.J.G.W. (1993) Management of bulking sludge control by the selector. In *Prevention and control of bulking activated sludge* (ed. D. Jenkins, R. Ramadori and L. Cingolani), pp.77–97. Centro Luigi Bazzucchi, Perugia.

Rheinheimer, G., Hegemann, W., Raff, J. and Sekoulov, I. (1988) *Stickstoffkreislauf im Wasser.* R. Oldenbourg Verlag, München.

Richard, M. (1989) *Activated sludge microbiology*. The Water Pollution Control Federation, Alexandria, VA.

Richard, M.G., Shimizu, G.P. and Jenkins, D. (1985) The growth physiology of the filamentous organism Type 021N and its significance to activated sludge bulking. *J. Wat. Pollut. Control Fed.* **57** (12), 1152–1162.

Sanden, B., Grunditz, C., Hansson, Y. and Dalhammar, G. (1994) Quantification and characterization of *Nitrosomonas* and *Nitrobacter* using monoclonal antibodies. *Wat. Sci. Technol.* **29** (7), 1–6.

Schade, M. and Lemmer, H. (1994) Counting bacteria of selected metabolic groups in activated sludge. *Wat. Sci. Technol.* **29** (7), 75–79.

Schlegel, H.G. (1985) *Allgemeine Mikrobiologie*. Georg Thieme Verlag, Stuttgart.

Seviour, E.M., Williams, C.J., Seviour, R.J., Soddell, J.A. and Lindrea, K.C. (1990) A survey of filamentous bacterial populations from foaming activated sludge plants in eastern states of Australia. *Wat. Res.*. **24** (4), 493–498.

Sezgin, M., Jenkins, D. and Parker, D.S. (1978) A unified theory of filamentous activated sludge bulking. *J. Wat. Pollut. Control Fed.* **50** (2), 362–381.

Shimizu, G.P. (1985) The growth and nutrient uptake kinetics of Type 021N, A novel *Thiothrix*-like bacterium responsible for activated sludge bulking. Ph.D. dissertation, University of California, Berkeley.

Slijkhuis, H. and Deinema, M.H. (1982) The physiology of *Microthrix parvicella*, a filamentous bacterium isolated from activated sludge. In *Bulking of activated sludge: preventative and remedial methods* (ed. B. Chambers and E.J. Tomlinson), pp. 75–89. Ellis Horwood, Chichester.

Soddell, J.A., Seviour, R.J., Seviour, E.M. and Stratton, H.M. (1993) Foaming and foam control in activated sludge systems. In *Prevention and control of bulking activated sludge* (ed. D. Jenkins, R. Ramadori and L. Cingolani) pp. 115–132. Luigi Bazzucchi Center, Perugia.

Sorm, R., Bortone, G., Saltarelli, R., Jenicek, P., Wanner, J. and Tilche, A. (1995) Phosphate uptake under anoxic conditions and fixed-film nitrification in nutrient removal activated sludge system. *Wat. Res.* (submitted).

Stensel, H.D. and Barnard, J.L. (1992) Principles of biological nutrient removal. In *Design and retrofit of wastewater treatment plants for biological nutrient removal.* (ed. C.W. Randall *et al.*), pp. 25–84. Technomic Publishing Co., Lancaster, PA.

Sudo, R. and Aiba, S. (1984) Role and function of protozoa in the biological treatment of polluted waters. In *Advances in biochemical engineering/biotechnology*, vol. 29 (ed. A. Fiechter), pp.117–141. Springer-Verlag, Berlin.

Tandoi, V. and Ramadori, R. (1991) Biochemical and microbiological aspects in enhanced biological phosphorus removal processes. In *Biological Approach to Sewage Treatment Process: Current Status and Perspectives*, (ed. P. Madoni), pp. 353–359. Luigi Bazzuchi Center, Perugia, Italy.

Toerien, D.F., Gerber, A., Lötter, L.H. and Cloete, T.E. (1990) Enhanced biological phosphorus removal in activated sludge systems. In *Advances in microbial ecology*, vol. 11 (ed. K.C. Marshall), pp. 173–230. Plenum Press, New York.

van Niekerk, A.M., Jenkins, D. and Richard, M.G. (1987) The competitive growth of *Zoogloea ramigera* and Type 0021N in activated sludge and pure culture – a model for low F:M bulking. *J. Wat. Pollut. Control Fed.* **59** (5), 262–273.

Verhagen, F.J.M. and Laanbroek, H.J. (1992) Effects of grazing by flagellates on competition for ammonium between nitrifying and heterotrophic bacteria in chemostats. *Appl. Envir. Microbiol.* **58** (6), 1962–1969.

Vlekke, G.J.F.M., Comeau, Y. and Oldham, W.K. (1988) Biological phosphate removal from wastewaters with oxygen or nitrate in sequencing batch reactors. *Envir. Technol. Lett.* **9**, 791–796.

Wagner, M., Amann, R., Lemmer, H., Manz, W. and Schleifer, K.H. (1994) Probing activated sludge with fluorescently labeled rRNA targeted oligonucleotides. *Wat. Sci. Technol.* **29** (7), 15–23.

Wanner, J. (1991) Use of the routine biological analyses to check the performance of sewage treatment plants. In *Biological Approach to Sewage Treatment Process: Current Status and Perspectives* (ed. P. Madoni), pp. 199–210. Luigi Bazzuchi Center, Perugia, Italy.

Wanner, J. (1992) Comparison of biocenoses from continuous and sequencing batch reactor. *Wat. Sci. Technol.* **25** (6), 239–249.

Wanner, J. (1994) *Activated sludge bulking and foaming control*. Technomic Publishing Co., Lancaster, PA.

Wanner, J. and Roskota, J. (1993) Metabolic selection in full-scale plant. *Newsletter of the IAWQ Specialist Group on Activated Sludge Population Dynamics* **5** (1), 4–8.

Wanner, J. and Grau, P. (1989) Identification of filamentous microorganisms from activated sludge: a compromise between wishes, needs and possibilities. *Wat. Res.* **23** (7), 883–891.

Wanner, J., Cech, J.S. and Kos, M. (1992) New process design for biological nutrient removal. *Wat. Sci. Technol.* **25** (4/5), 445–448.

Williams, T.M. and Unz, R.F. (1985a) Isolation and characterization of filamentous bacteria present in bulking activated sludge. *Appl. Microbiol. Biotechnol.* **22**, 273–282.

Williams, T.M. and Unz, R.F. (1985b) Filamentous sulfur bacteria of activated sludge: characterization of *Thiothrix*, *Beggiatoa*, and Eikelboom Type 021N strains. *Appl. Envir. Microbiol.* **49** (4), 887–898.

Williams, T.M., Unz, R.F. and Doman, J.T. (1987) Ultrastructure of *Thiothrix* spp. and Type 021N bacteria. *Appl. Envir. Microbiol.* **53** (7), 1560–1570.

Ziegler, M., Lange, M., Kampfer, P., Hoffmeister, D., Weltin, D., Winkler, B. and Dott, W. (1988) Occurrence of filamentous bacteria in activated sludge (bulking sludge), isolation and characterization. *Wat. Sci. Technol.* **20** (11/12), 497–499.

4. Molecular techniques for determining microbial community structures in activated sludge

Michael Wagner and Rudolf Amann

4.1 Comparison between cultivation-dependent methods and community analyses *in situ*

The activated sludge treatment of sewage is probably today's most important biotechnological process. In the past decades great progress has been achieved in process engineering, but our current knowledge on the structures and dynamics of the involved microbial communities, and consequently our understanding of the microbiology of the activated sludge process, is still very limited. These microbial consortia have mostly been analysed by culture-dependent methods such as viable plate count or most probable number (MPN) techniques (Dias and Bhat 1964; Prakasam and Dondero 1967; Benedict and Carlson 1971). However, for their inevitable selectivity towards certain organisms these methods represent no adequate means of addressing this question (Wagner *et al.* 1993). Standard plate counts reflect not so much the actual bacterial community structure of the activated sludge but rather the selectivity of growth media for certain bacteria. For oligotrophic to mesotrophic aquatic habitats, direct microscopic counts can exceed viable cell counts by several (2–4) orders of magnitude. In eutrophic water this effect is less pronounced, but recoveries from activated sludge even with optimized media are only between 0.85% (Pike and Curds 1971) and 14% (Wagner *et al.* 1993). This phenomenon has been known to microbiologists for generations. By now there is no doubt that in most cases most of the microscopically visible cells are viable, but do not form visible colonies on plates. In activated sludge this is mainly caused by two effects:

(1) clumping of the cells in the activated sludge flocs prevents a quantitative release of individual bacteria and consequently leads to an underestimation of the number of active cells determined by viable plate count techniques
(2) the cultivation conditions are not suitable for all bacteria; this might apply both to well-known strains and to hitherto uncultured bacteria.

Even after decades of thorough microbiological studies it is possible that the species described for activated sludge represent only the minor part of the true microbial diversity in this environment. Restrictions and potential biases in the cultivation techniques should motivate microbiologists to perform more studies *in situ*.

A possible way to overcome the quantitative and qualitative biases introduced by cultivation-dependent methods is the use of chemotaxonomic techniques for direct examination of activated sludge. Hiraishi *et al.* (1988, 1989) utilized bacterial respiratory quinone profiles to monitor different bacterial populations in activated sludge. Auling and co-workers (1991) determined the polyamine composition of anaerobic–aerobic activated sludge systems for an estimation of the microbial population structure. In the same study the polyamine diaminopropane was employed as a specific biomarker for bacteria of the genus *Acinetobacter*. However, studies with a chemotaxonomic approach in activated sludge have been scarce. This is probably due to some intrinsic limitations. First, statements about the specificity of respiratory quinone or polyamine profiles are based on a limited number of examined, culturable bacteria. Thus the recent detection of diaminopropane in bacteria of the genus *Aeromonas* (P. Kämpfer, personal communication) demonstrated the restricted validity of diaminopropane as a biomarker for the genus *Acinetobacter*. Secondly, given the complexity of activated sludge the relative quantities of particular biomarkers can hardly be converted into accurate estimates of cell numbers. The direct identification and enumeration of bacteria in activated sludge require different techniques that analyse at the cellular level. One possibility is the immunofluorescence approach (Bohlool and Schmidt 1980), which has been used successfully to monitor bacteria in complex environments. In detail, this technique has been applied to identify *Sphaerotilus natans* (Howgrave-Graham and Steyn 1988), *Acinetobacter* spp. (Cloete and Steyn 1988), *Nitrosomonas* spp. (Völsch *et al.* 1990), *Legionella* spp. (Palmer *et al.* 1993) and *Nocardia amarae* (Hernandez *et al.* 1993) in activated sludge. However, the technique using fluorescent antibodies can be disturbed by extracellular polymeric substances that have been reported to hinder antibody penetration (Szwerinski *et al.* 1985). In addition, non-specific binding of antibodies to detritus particles and fungal spores can result in high levels of background fluorescence. In addition, the production of specific antibodies requires a pure culture of the organism of interest. Keeping in mind that most bacteria in activated sludge might be still uncultured, new tools for the identification of bacteria *in situ* are necessary that are completely independent of cultivation methods. It should, however, be stressed that owing to its high specificity immunofluorescence will remain a valid and powerful method for studying the autecology of specific populations in activated sludge.

4.2 The rRNA approach to community analysis

Today, a combination of direct retrieval of rRNA sequences by the polymerase chain reaction (PCR) and fluorescent probing *in situ* enables an identification and phylogenetic characterization of microorganisms without cultivation (Amann *et al.* 1991; Spring *et al.* 1992). As outlined in Figure 4.1a, one can start with the extraction of DNA from an environmental sample, use standard molecular techniques to obtain a clone library and to retrieve rDNA sequence information, and perform a comparative analysis of the retrieved sequences. This yields information on the identity or relatedness of new sequences by comparison with the available databases and gives a minimal estimate of the genetic diversity in the examined sample. However, this does not prove that the sequences were retrieved from cells thriving in this habitat. They can also originate from naked DNA present in the sample, or from contaminants. Therefore, in the second phase, sequence-specific hybridization probes have to be designed to identify and enumerate whole fixed cells in the original sample by hybridization *in situ*. Often it is also interesting to quantify the abundance of a certain rRNA or rDNA in the extracted nucleic acid pool by dot-blot hybridization or of a certain clone in the library by colony hybridization. Only the combination of probing and sequencing allows the full exploitation of the potential of this approach.

4.2.1 Sequencing

First attempts to characterize environmental samples by studying rRNA started about a decade ago. In these studies 5 S rRNA molecules were directly extracted from mixed samples, the molecules belonging to the different community members were separated electrophoretically, and a comparative sequence analysis yielded phylogenetic placings (Stahl *et al.* 1984, 1985; Lane *et al.* 1985). These pioneering studies yielded interesting insights. However, the information content of the *ca.* 120 nucleotide long 5 S rRNA is relatively small and the requirement for an electrophoretic separation of the different 5 S rRNA molecules limited this approach to less complex ecosystems. Consequently, the use of the larger rRNA molecules for studies in microbial ecology has been suggested (Olsen *et al.* 1986). An average bacterial 16 S rRNA molecule has a length of 1,500 nucleotides; 23 S rRNA molecules are around 3,000 nucleotides. Both contain more than sufficient information for reliable phylogenetic analyses. The principal steps of the proposed procedure were: (1) extraction of total community DNA, (2) preparation of a shotgun DNA library in bacteriophage lambda, (3) screening by hybridization with a 16 S rRNA-specific probe, (4) sequence determination from clones containing 16 S rRNA genes and (5) comparative analysis of the retrieved sequences (see Figure 4.1). The first thorough application of this approach was the characterization of marine picoplankton (Schmidt *et al.* 1991). Many unknown sequences could be identified in this library.

With the advent of the PCR (Saiki *et al.* 1988) a method became available to speed up this quite laborious procedure. By PCR 16 S rRNA gene fragments can be selectively amplified from mixed DNA. Gene libraries derived from mixed amplificates should contain only defined fragments that can be rapidly sequenced from known priming sites. This approach greatly reduces or even obviates lengthy screening procedures, which were necessary to identify the rRNA containing clones in shotgun libraries. However, it also bears an additional potential bias for the representative assessment of the natural abundance of rRNA genes, the preferential amplification of certain templates. Owing to selective priming or higher-order structure elements, certain sequences might be discriminated against slightly in comparison with others. This would result in huge differences after multiple cycles. It is also important to check for contamination, a problem commonly encountered with PCR.

Finally, there is yet another route to the molecular characterization of natural microbial communities. Cloning of cDNA transcribed from 16 S rRNA with the enzyme reverse transcriptase allows us, as with PCR, to retrieve useful sequence information selectively (Weller and Ward 1989; Ward *et al.* 1990). A variation of this approach involves a PCR step with rDNA-specific primers immediately after the reverse transcription (Amann *et al.* 1992). Starting the sequence retrieval from rRNA instead of DNA has the advantage that because of the smaller size of rRNA more rigorous nucleic acid extraction techniques (e.g. bead beating (Stahl *et al.* 1988)) can be applied.

Only hitherto unknown sequences were retrieved in all the initial experiments. This clearly indicated that the sequences present in the 16 S rRNA databases (Larsen *et al.* 1993, Neefs *et al.* 1993) – mostly culture collection strains, currently encompassing about 35% of the validly described species – are not representative of the naturally occurring diversity of microorganisms.

4.2.2 Probe design and evaluation

Biases in the rRNA sequence retrieval can occur at the level of DNA extraction, amplification and also in the cloning step. Furthermore quantification of the relative abundances of certain populations from the relative abundances of certain rDNA clones will be biased by the fact that the number of rRNA gene operons present in a bacterial chromosome can range from one to ten copies.

It is therefore important to corroborate the presence and to estimate the abundance of the retrieved sequences, whether within the initially extracted nucleic acids, the mixed pool of amplificates or preferably within the cells present in the original environmental sample, as indicated in Figure 4.1. This can best be achieved by hybridization with nucleic acid probes. The design of such probes has been reviewed in detail (Stahl and Amann 1991). The principal steps are the alignment of rRNA (gene) sequences, the identification of sequence idiosyncrasies, the synthesis and labelling of complementary nucleic acid probes, and finally the experimental evaluation and optimization of the probe specificities and assay sensitivities. Our knowledge of the microorganisms and sequences present in a certain ecosystem is obviously limited by the number of sequences that have been retrieved and determined. A probe, intended to be specific for a unique sequence, has to be compared with all accessible rRNA sequences. It also has to be evaluated by hybridization against cultured reference strains (Devereux *et al.* 1992, Manz *et al.* 1992). The thorough knowledge of the 16 S and 23 S rRNA conservation profiles does also allow the investigator to tailor a probe complementary to larger assemblages. This can be advantageous if the goal of a study is not to follow the fate of a narrowly defined taxonomic unit but to get an overview on the community structure and the dynamics of a complex ecosystem.

4.2.3 Quantitative dot-blot hybridization

A quantification of a certain 16 S rRNA compared with total 16 S rRNA can be obtained by dot-blot hybridizations of a directly isolated nucleic acid mixture with universal and specific oligonucleotide probes. The relative abundance is calculated by dividing the amount of specific probe bound to a given sample by the amount of hybridized universal probe (measured, for example, as counts per minute for radioactively labelled probes). This approach was first applied to the monitoring of population changes in the rumen of cattle (Stahl *et al.* 1988). By using radioactively labelled oligonucleotide probes in a dot-blot assay, rRNA sequences with relatively low abundances between 0.1% and 1% could be quantified. One should be aware that, as for the chemotaxonomic markers, relative rRNA abundance cannot be directly translated into cell numbers: cells of different species have different ribosome contents, ranging roughly from 10^3 to 10^5 ribosomes per cell, and even for one strain the cellular rRNA contents can vary significantly (at least over one order of magnitude) because they are directly correlated with the growth rate (Schaechter *et al.* 1958).

The relative rRNA abundance should, however, represent a reasonable measurement of the relative physiological activity of the respective population because it is the product of the number of detected cells and the average rRNA content. This information on the general activity of a given population should not automatically be regarded as an indication of a specific kind of activity. Often one population has the potential to catalyse different transformations. One genotype might be linked to several phenotypes.

4.2.4 Probing *in situ* with rRNA-targeted oligonucleotide probes

The use of hybridization *in situ* for counting and identifying organisms was proposed by Olsen *et al.* (1986). As demonstrated by the immunofluorescence approach, fluorescent probes yield superb spatial resolution and can be detected instantaneously by epifluorescence microscopy. Fluorescently monolabelled, rRNA-targeted oligonucleotide probes were shown to allow the detection of individual cells (DeLong *et al.* 1989). This made whole-cell hybridization with rRNA-targeted probes a suitable tool for determinative, phylogenetic and environmental studies in microbiology (Amann *et al.* 1990b). Like immunofluorescence, whole-cell hybridization with fluorescently labelled, rRNA-targeted oligonucleotides can be combined with flow cytometry for a high-resolution automated analysis of mixed microbial populations (Amann *et al.* 1990a, Wallner *et al.* 1993). Analysing at a single-cell level, whole-cell hybridization can obviously provide a more detailed picture than dot-blot hybridization. One can not only determine the cell morphology of an uncultured microbe and its abundance, but also

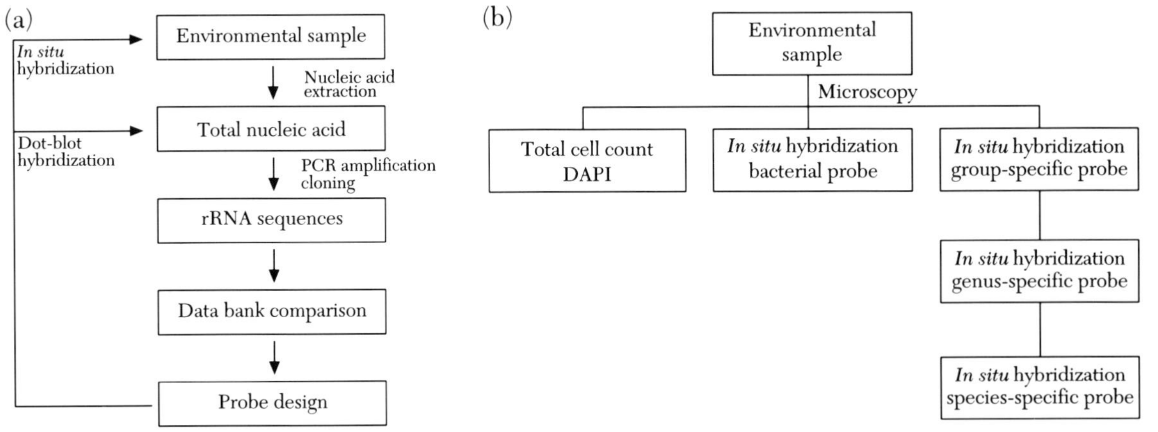

Figure 4.1. (a) General flow chart for the rRNA approach to community analyses; (b) general flow sheet for the top-to-bottom approach to community analyses.

analyse spatial distributions *in situ*. Quantification of the signal conferred by rRNA-targeted oligonucleotides should even allow the estimation of growth rates of individual cells *in situ* (Poulsen *et al.* 1993).

4.2.5 Application of group-specific oligonucleotide probes

The rRNA approach to the phylogenetic identification and detection of microorganisms *in situ* is often used in such a way that the samples are first characterized by the retrieval of rDNA sequences and then, in a second phase, the abundance of certain clones, specific rRNA sequences or sequence-harbouring cells is quantified with oligonucleotide probes in a colony, dot-blot or whole-cell hybridization. Although PCR-assisted approaches are very helpful, the analysis of complex communities remains rather laborious and time-consuming, forcing most researchers to cut down on the number of examined clones. Even by examining several hundred clones smaller populations (around or below 1%) might not be detected. Therefore there is a need for an alternative path to a direct molecular characterization of environmental samples. It has been well established that on various taxonomic levels phylogenetic groups are characterized by rRNA sequence idiosyncrasies, so-called **signatures**. New isolates can be allocated to, for example, one of the major bacterial taxa by signature analysis (Woese *et al.* 1985a). Oligonucleotide probes are more suitable for routine screening of such signatures than sequencing (Giovannoni *et al.* 1988; Stahl *et al.* 1988). Such probes have been designed for the highest taxonomic levels, the domains Archaea, Bacteria and Eukarya (Amann *et al.* 1990b; DeLong *et al.* 1989; Giovannoni *et al.* 1988), for intermediate levels, for example for the two groups of Archaea, the Crenarchaeota and Euryarchaeota (Burggraf *et al.* 1997), for the Gram-negative, sulphate-reducing bacteria (Amann *et al.* 1992), for the α, β and γ subclasses of Proteobacteria (Manz *et al.* 1992), for clusters within the 'Flavobacter-Bacteroides' phylum (Gherna and Woese 1992; W. Manz, R. Amann, M. Vancanneyt and K.-H. Schleifer, in preparation), for Gram-positives with a high G+C DNA content (Roller *et al.* 1997) and for lower taxa, e.g. genus-, species-, and subspecies-specific probes (see, for example, Amann *et al.* 1992; Devereux *et al.* 1992; Goebel *et al.* 1987; Wagner *et al.* 1994, 1997). If such probes are applied to parallel subsamples in an ordered top-to-bottom approach, initially using the domain-specific probes followed by probes of narrower and narrower specificity, increasingly refined information on the community diversity and composition can be obtained very rapidly by both dot-blot and whole-cell hybridization (Figure 1b). In this 'nested' approach (Stahl 1986) the information gained with the higher-level probes are used to select the probe sets for the next lower taxonomic level; for example, if the group-specific probing reveals a dominance of bacteria belonging to the γ subclass of *Proteobacteria*, probes specific for the genuine pseudomonads (Schleifer *et al.* 1992), the *Enterobacteriaceae* (N. Springer, W. Ludwig, R. Amann and K.-H. Schleifer, unpublished) and the genera *Acinetobacter* (Wagner *et al.* 1994) and *Aeromonas* (P. Kämpfer, M. Wagner, R. Amann, R. Erhart, C. Beimfohr and K.-H. Schleifer, in preparation) could be applied in the next step. One has to be aware that probes are designed based on a large though limited database. Their specificity is evaluated by hybridization against culturable reference organisms (Devereux *et al.* 1992, Manz *et al.* 1992). This is an obvious drawback because the currently culturable microorganisms represent only the minor fraction of the true diversity of organisms and sequences. However, the situation will improve with the integration of more (for example, also the directly retrieved) rDNA sequences in the public databases. This will result in refined infor-

Table 1. EUB/DAPI ratio in different sewage treatment plants

Sewage treatment plant	EUB/DAPI ratio (%)	Reference
München I, aeration basin	89 ± 7	Manz *et al.* (1994)
München II, high load aeration basin, stage 1	89 ± 7	Wagner *et al.* (1993)
München II, low load aeration basin, stage 2	70 ± 7	Wagner *et al.* (1993)
Hirblingen, anaerobic stage	78 ± 15	Wagner *et al.* (1994)
Hirblingen, anoxic stage	82 ± 11	Wagner *et al.* (1994)
Hirblingen, aerobic stage	79 ± 11	Wagner *et al.* (1994)
Aretsried, sequencing batch reactor	89 ± 9	Manz *et al.* (1994)

mation on the 'true' specificity of old probes and in the design of new probes. rRNA-targeted oligonucleotide probes should be regarded as tools being subject to refinement.

4.3 Probing activated sludge with rRNA-targeted oligonucleotide probes

Although cultivation-dependent methods record only a small percentage of the total viable bacteria, they demonstrate that activated sludge samples contain a whole array of different microbial species. As mentioned above, a characterization of such complex microbial communities by retrieval of rDNA sequences and subsequent oligonucleotide probing would probably yield a vast collection of different rDNA sequences and specific oligonucleotide probes. A thorough investigation would be very time-consuming and tedious, and would suffer from a lack of orientation regarding the 'importance' of the retrieved sequences. As an alternative, probing for important phylogenetic groups in a top-to-bottom approach allows a more directed and rapid, yet less detailed, characterization of the microbial community structure.

4.3.1 The EUB/DAPI ratio

The applicability of fluorescent oligonucleotide probing to activated sludge samples can be tested with the Bacteria-specific probe EUB. Dual staining of activated sludge samples with probe EUB and the DNA intercalating dye 4,6-diamidino-2-phenylindole (DAPI) (Hicks *et al.* 1992) reveals that in all sewage treatment plants hitherto examined, most of the microbial cells revealed by intercalation of the DNA-specific dye DAPI also emitted probe-conferred fluorescence (Table 4.1).

The fact that *ca.* 80% of total cells hybridized with the bacterial probe proved that the bulk of cells present in the activated sludge samples were bacteria that had sufficient rRNA for detection and were permeabilized for oligonucleotide probes by the standard fixation procedures. The 'high EUB/DAPI ratio' indicated that most bacteria in activated sludge are growing and are metabolically active. Bacteria stained with DAPI that do not hybridize with the probe EUB are either metabolically inactive or not permeable for oligonucleotide probes by standard fixation procedures. Figure 4.2 shows a comparison of the EUB/DAPI ratio with total viable counts on different media for an activated sludge sample from the sewage treatment plant München I. Whereas 76% of all bacteria can be detected by hybridization *in situ* with probe EUB, only 3–19% of total bacteria can be cultivated even on optimized media. Therefore, in contrast with cultivation-dependent methods, hybridization techniques *in situ* have the potential to obtain a more complete view of the diversity and dynamics of the microbial consortia involved in the activated sludge process.

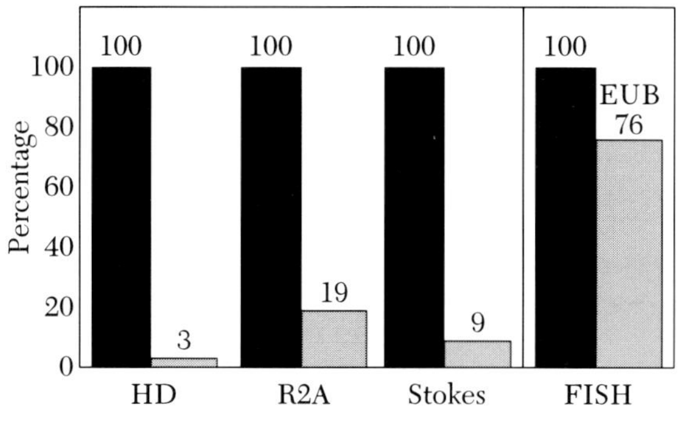

Figure 4.2. Comparison of the EUB/DAPI ratio with total viable counts on three different media for an activated sludge sample taken from the sewage treatment plant München I. Solid bars, total cell counts; grey bars, CFU. FISH, fluorescent in situ hybridization.

4.3.2 Differentiation between the subclasses of Proteobacteria

In the class Proteobacteria (Stackebrandt *et al.* 1988) we find most of the classical 'Gram-negative bacteria'. According to molecular data this class was further subdivided into four subclasses, named α to δ (Woese 1987). These subclasses comprise many of the well-known culturable wastewater bacteria:

(1) among others the α subclass encompasses *Caulobacter* spp., *Nitrobacter* spp. and some *Zoogloea* spp.
(2) members of the genera *Alcaligenes* and

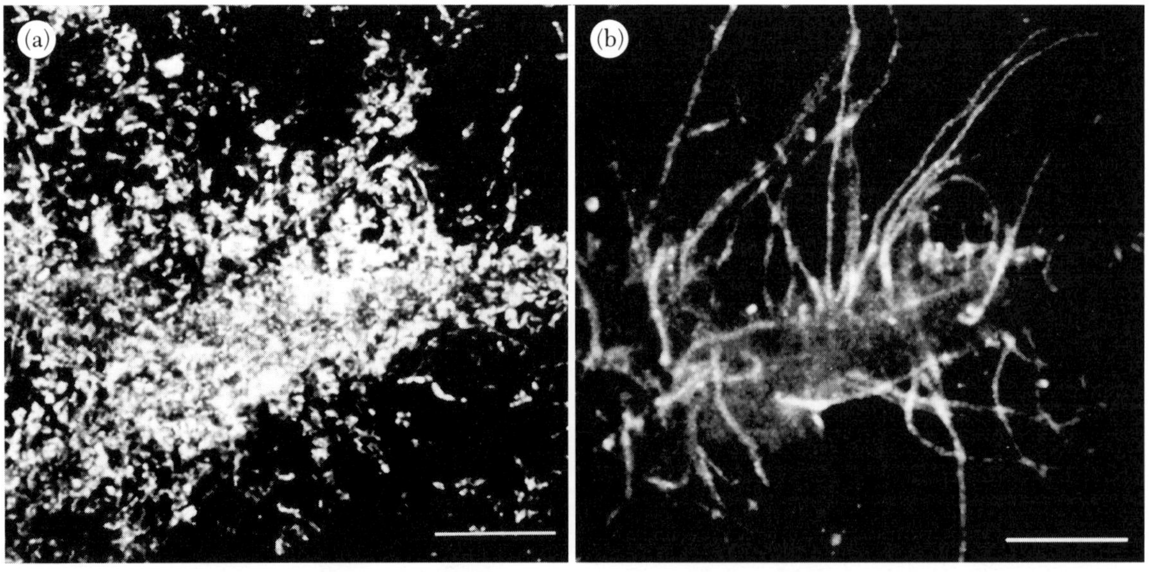

Figure 4.3. Two-dimensional reconstructions by SCLM-analyses of an activated sludge sample obtained from the sewage treatment plant München I after simultaneous hybridization in situ with fluorescein-labelled probe BET (a) and tetramethylrhodamine-labeled probe GAM (b). Scale bars, 20 μm. Background fluorescence of the floc material was minimized by confocal microscopy.

Comamonas, Sphaerotilus spp., some *Zoogloea* spp. and most of the autotrophic nitrifiers are members of the β subclass

(3) among others, *Acinetobacter* spp., *Aeromonas* spp, Enterobacteriaceae, fluorescent pseudomonads and *Vibrio* spp. belong to the γ subclass

(4) The δ subclass encompasses sulphate-reducing bacteria and myxobacteria.

Ramsing *et al.* (1993) investigated the vertical distribution of sulphate-reducing bacteria (δ subclass of Proteobacteria) in photosynthetic biofilms from a trickling filter with hybridization *in situ* with probe SRB (Amann *et al.* 1992). A negative correlation between the vertical distribution of detected cells and O_2 profiles (measured by microelectrodes) was demonstrated. Manz *et al.* (1992) described oligonucleotide probes specific for the α, β and γ subclasses of Proteobacteria. Probing activated sludge with fluorescent derivatives of these oligonucleotides has revealed that the microbial consortia of various municipal treatment plants are dominated by Proteobacteria (see, for example, Wagner *et al.* 1993). Members of the β subclass play the major role of the microbial consortia in municipal activated sludge. Bacteria of the α and γ subclasses are less abundant (P. Kämpfer, M. Wagner, R. Amann, R. Erhart, C. Beimfohr and K.-H. Schleifer, in preparation; Manz *et al.* 1994; Wagner *et al.* 1993, 1994). Figure 4.3 shows a representative dual-label hybridization experiment of an activated sludge sample with probes specific for the β and γ subclasses of Proteobacteria. Figure 4.4a displays the *in situ* distribution among the α, β and γ subclasses in an activated sludge sample from the sewage treatment plant München I: 30% of the bacterial cells detectable with probe EUB could be assigned to the β subclass of Proteobacteria (probe BET); 17% of the cells were members of the γ subclass of Proteobacteria (probe GAM); bacteria of the α subclass of Proteobacteria represent 12% (probe ALF). The community composition *in situ* differed markedly from the heterotrophic saprophyte index obtained by cultivation on three different media. The nutrient-rich medium HD (Figure 4.4b) resulted in an overestimation of members of the γ subclass of Proteobacteria (55%) and selected against the α (3%) and β subclasses (8%) of Proteobacteria. This 'γ-shift' in response to cultivation on nutrient-rich media could also be observed in other activated sludge samples from municipal sewage treatment plants (Wagner *et al.* 1993, 1994; Manz *et al.* 1994) and a dairy wastewater plant (Manz *et al.* 1994). With the low-nutrient medium R_2A (Figure 4.4c), recommended for the enumeration and subculturing of bacteria from potable water (Reasoner and Geldreich 1985), no α-subclass Proteobacteria could be retrieved, whereas the relative abundance of colonies of bacteria of the β (29%) and γ (15%) subclasses was comparable to the abundance *in situ*. Plating on slightly modified Stokes medium (Figure 4.4d), recommended for the cultivation of the β subclass of the proteobacterium *Sphaerotilus natans* (Rouf and Stokes 1964), resulted, as expected, in a large number of β-subclass Proteobacteria (48%). In comparison with the community structure *in situ*, the proportion of bacteria of the α (3%) and γ subclasses (9%) was underestimated.

Because none of the tested media yielded CFUs above 19% of the total cells, and all media produced cultivation-dependent shifts in community composition, analysis *in situ* is required for obtaining realistic values. As β-subclass Proteobacteria numerically predominate in various municipal sewage treatment plants, future studies trying to isolate and sequence 16 S or 23 S rDNA of members of this group from activated sludge will be important to broaden our knowledge of this obviously important group of wastewater bacteria.

4.3.3 Other important groups: 'Flavobacter-Bacteroides' phylum, Gram-positives with a high G+C DNA content

Cytophagae and flavobacteria have been isolated regularly from biological sewage treatment facilities (Güde 1980). Also, the well-known Gram-negative filamentous bacterium *Haliscomenobacter hydrossis*, frequently encountered in activated sludge samples, is a member of the 'Flavobacter-Bacteroides' phylum (Gherna and Woese 1992). Cultivation of bacteria belonging to the 'Flavobacter-Bacteroides' phylum from activated sludge can require 10–24 days (Güde 1980). For monitoring this phylum *in situ*, Manz *et al.* (manuscript in preparation) developed a 16 S rRNA-targeted oligonucleotide probe (CF) specific for many species of the genera *Cytophaga*, *Flavobacterium* and *Saprospira*. Probing *in situ* of activated sludge originating from two municipal sewage treatment plants revealed that 8–23% of all bacterial cells could be detected simultaneously with probe CF (Wagner *et al.* 1994; Manz *et al.* 1994). The microbial community of a dairy wastewater plant was shown to be even numerically dominated by spindle-shaped bacteria belonging to the 'Flavobacter-Bacteroides' phylum (Manz *et al.* 1994). In these studies bacterial cells hybridizing with probe CF have been most frequently detected in the cores of the flocs. Figure 4.4 shows the inadequacy of three different cultivation media to reflect accurately the real numerical occurrence of representatives of this important taxon in activated sludge.

Molecular taxonomists found significant rRNA homologies between actinomycetes and a group of related bacteria, all characterized by Gram-positive staining and a high G+C DNA content, and consequently created a group named 'actinomycetes branch' (Stackebrandt 1991) or 'Gram-positives with high G+C content of DNA' (Woese 1987). Many species in this phylogenetic group have biotechnological or medical importance. Rapid and reliable monitoring of Gram-positive bacteria with a high G+C DNA content in activated sludge is necessary because representatives of the 'actinomycetes branch' have frequently been ascribed an important role in sewage purification processes (Adamse 1968; Seiler *et al.* 1980). For instance, actinomycetes have been connected with malfunctions such as foaming (Soddell and Seviour 1990; Blackall *et al.* 1991). Roller *et al.* (1997) constructed a 23 S rRNA-targeted

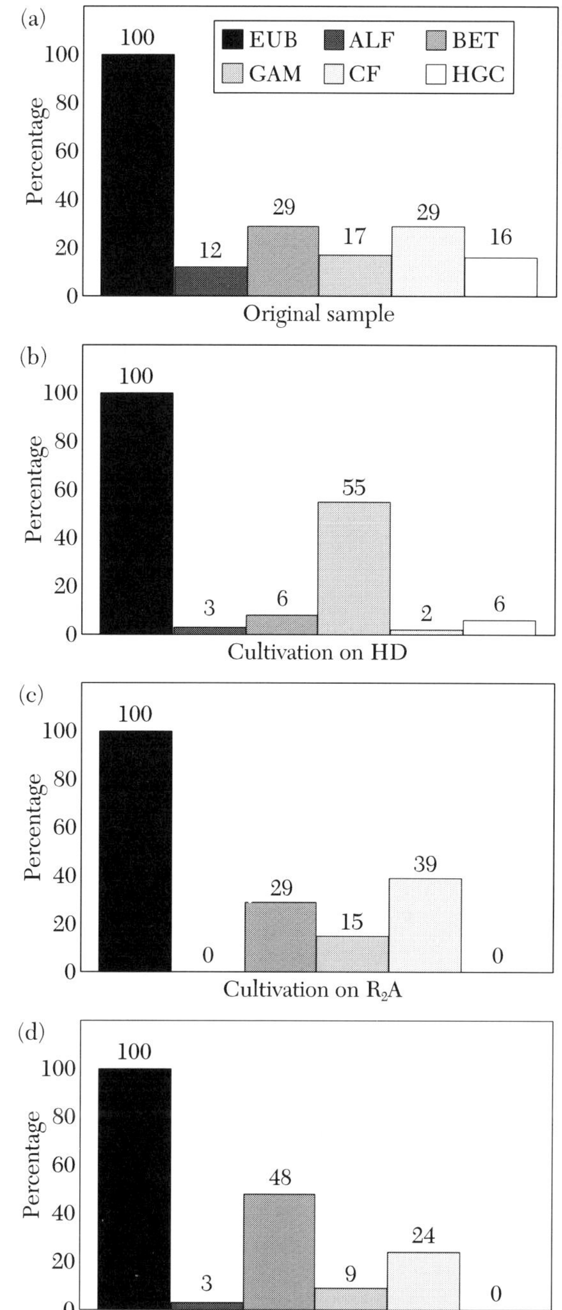

Figure 4.4. Comparison of community compositions in an activated sludge sample (identical sample to that in figure 4.2). (a) Microbial community structure, as determined by hybridization in situ. (b) Heterotrophic saprophytes, as determined by whole-cell hybridization of isolates obtained on HD from the same sample. (c) Heterotrophic saprophytes, as determined by whole-cell hybridization of isolates obtained on R_2A from the same sample. (d) Heterotrophic saprophytes, as determined by whole-cell hybridization of isolates obtained on Stokes from the same sample.

group-specific oligonucleotide probe HGC for this important taxon. About one-quarter of all bacteria in an activated sludge sample taken from a plant treating dairy waste could be specifically visualized by hybridization *in situ* with probe HGC (Manz *et al.* 1994). In the same study only 9% of all bacteria in an activated sludge sample originating from a municipal sewage treatment plant with chemical phosphate precipitation were identified as Gram-positive bacteria with a high G+C DNA content. Analyses *in situ* of activated sludge samples from municipal sewage treatment plants with enhanced biological phosphate removal (EBPR) with the group-specific probe HGC have demonstrated a high abundance of members of the 'actinomycetes branch' (Wagner *et al.* 1994). This would not have been detected by standard plating because the examined cultivation media selected against Gram-positive bacteria with a high G+C DNA content (Figure 4.4).

4.3.4 Probing at the genus or species level

4.3.4.1 The genus Acinetobacter

According to 16 S rRNA sequence analysis, the genus *Acinetobacter* belongs to the γ subclass of *Proteobacteria* (Woese *et al.* 1985b). Acinetobacters are frequently found in soil, water and sewage (Towner 1992). In activated sludge their occurrence has been linked to enhanced biological phosphate removal (EBPR) (see, for example, Fuhs and Chen 1975; Streichan *et al.* 1990). However, there is a considerable discrepancy between the numbers of acinetobacters isolated from activated sludge by conventional plate count methods and estimations inferred from direct examinations by chemotaxonomic and immunochemical techniques. On the basis of conventional cultivation methods, acinetobacters represent a major part of the bacteria present in EBPR systems (see, for example, Buchan 1983; Brodisch 1985; Lötter and Murphy 1985), whereas chemotaxonomic and immunochemical data indicate that they may be present only in rather low numbers (Cloete and Steyn 1988; Hiraishi *et al.* 1989; Auling *et al.* 1991). A 16 S rRNA-targeted oligonucleotide probe specific for the genus *Acinetobacter* (Wagner *et al.* 1994) was used for monitoring of acinetobacters *in situ* in the anaerobic and aerobic compartments of a sewage treatment plant with EBPR (Figure 4.5b). In the same study microbial community structures were further analysed with group-specific probes. In the anaerobic and aerobic basins the activated sludge was dominated by members of the β subclass of Proteobacteria and by Gram-positive bacteria with a high G+C DNA content. *Acinetobacter* spp. have been shown to comprise (usually significantly) less than 10% of the cells. Comparison of the microbial community structures *in situ* with the compositions of the heterotrophic saprophyte flora (determined by plating) revealed that the number of acinetobacters was significantly overestimated by cultivation on nutrient-rich medium. The high *in situ* abundance of Gram-positive bacteria with a high G+C DNA content, which was increased further by the addition of acetic acid to the return sludge, and the presence of polyphosphate inclusions in these bacteria (Ohsumi *et al.* 1980; Lemmer and Baumann 1988, Nakamura *et al.* 1989; Wagner *et al.* 1994) suggested that members of this group and not the acinetobacters are in many cases the key mediators for EBPR. Sewage treatment plants with EBPR seem particularly prone to foaming (Wanner and Grau 1989). As foaming is often connected with Gram-positive bacteria with a high G+C DNA content (e.g. *Nocardia amarae*, *N. pinensis*, *Rhodococcus rhodochrous* and *Tsukamurella paurometabolum*) (Lemmer 1986; Blackall *et al.* 1989; Sodell and Seviour 1990) this solid separation problem may be caused by the same organisms responsible for the EBPR process.

4.3.4.2 Gram-negative filamentous bacteria

Gram-negative filamentous bacteria are commonly observed in activated sludge and contribute to poor settlement of activated sludge flocs in secondary sedimentation tanks, a problem referred to as **sludge bulking**. From time to time, filamentous bulking is a serious problem in 40–50% of all activated sludge plants (see, for example, Blackbeard *et al.* 1986). By using morphological characters and simple staining reactions, Eikelboom (1975) distinguished 26 types of filamentous bacteria in activated sludge samples. However, this identification key has general limitations. First, morphology and staining reactions of microbial cells can vary over a broad range depending on environmental conditions. Non-filamentous growth forms have been described for the filamentous bacteria *Haliscomenobacter hydrossis* (van Veen *et al.* 1973), *Sphaerotilus natans* (Waitz and Lackey 1959; Mulder and Deinema 1992) and *Microthrix parvicella* (Foot *et al.* 1992). For *Leptothrix discophora*, it was suggested that sheath-forming capacity is encoded on easily lost genetic elements such as plasmids (Emerson and Ghiorse 1992). Certain filamentous organisms, e.g. *Microthrix parvicella* (Foot *et al.* 1992) and Eikelboom Type 1863 (Wagner *et al.* 1994) can even show variable Gram stain reactions. Secondly, the filamentous sulphur bacteria *Thiothrix* spp., Eikelboom Type 021N, and *Leucothrix mucor* are hardly differentiated by their morphology (Nielsen 1984; Brock 1992). *Thiothrix* sp. filaments without sulphur granules cannot be distinguished from *Leucothrix mucor* (Brock 1992). Discrimination between Thiothrix sp./021N and *Leucothrix mucor* can

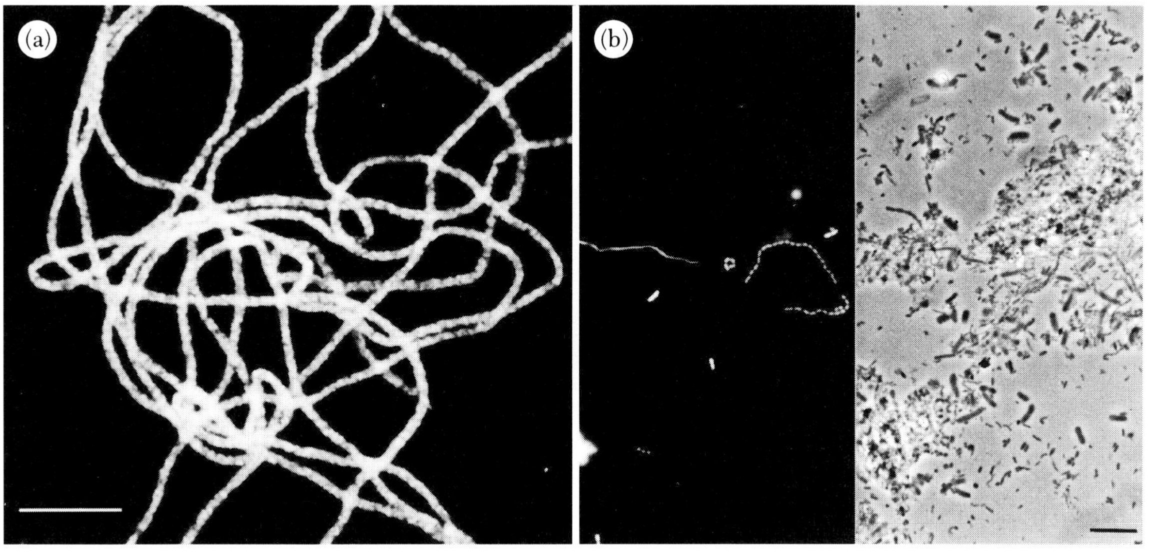

Figure 4.5. Hybridization in situ of activated sludge taken from the sewage treatment plant München I. (a) Two-dimensional reconstruction by SCLM analysis after hybridization with a tetramethylrhodamine-labelled probe specific for bacteria of Eikelboom Type 021N. Background fluorescence of the floc material was minimized by confocal microscopy. (b) Epifluorescence (left) and phase-contrast (right) micrographs are shown for an identical microscopic field. Hybridization in situ with a fluorescein-labelled probe specific for the genus Acinetobacter. Scale bars, 10 μm.

be achieved only by a quite time-consuming sulphur storage test (Nielsen 1984).

Given the scope of the problem and the fact that no reasonable monitoring is possible by cultivation techniques (Wagner *et al.* 1993), specific 16 S rRNA-targeted oligonucleotide probes have been developed for *Haliscomenobacter* spp., *Sphaerotilus* spp., *Leptothrix* spp., *Thiothrix* spp., *Leucothrix mucor* and bacteria of Eikelboom Type 021N (Wagner *et al.* 1997). These filamentous bacteria represent an important part of the Gram-negative filamentous bacteria in activated sludge (Farquhar and Boyle 1971; van Veen 1973; Eikelboom 1975; Nielsen 1984; Williams and Unz 1985; Jenkins *et al.* 1986; Mulder and Deinema 1992). Fluorescently labelled probe derivatives have been successfully applied for monitoring of these filamentous bacteria *in situ* in activated sludge samples of ten German sewage treatment plants. For example, filamentous bacteria of Eikelboom Type 021N could be visualized inside the sludge flocs (Figure 5a). For the microbiologist analysing the wastewater treatment process, hybridization *in situ* with probes specific for filamentous bacteria facilitates rapid and reliable identification of these bacteria in activated sludge. Understanding of the principles of filamentous bulking is the key to its control.

4.4 Analysing spatial distribution by scanning confocal laser microscopy

Probing activated sludge with fluorescently labelled oligonucleotide probes in combination with conventional epifluorescence microscopy not only allows a phylogenetic classification of bacteria without prior cultivation but also gives information about their spatial organization in the activated sludge flocs. However, the limited spatial resolution of conventional epifluorescence microscopy in thick samples as well as autofluorescence originating from mineral or organic particles results in images that lack clarity and fail to represent accurately the three-dimensional arrangement of the cells. The optical sectioning properties of a scanning confocal laser microscope (SCLM) allow one to avoid most of these problems and improve data retrieval (White *et al.* 1987; Lawrence *et al.* 1991). Combination of hybridization *in situ* with fluorescently labelled, rRNA-targeted oligonucleotide probes with SCLM (Embley and Finlay 1994; Wagner *et al.* 1997) considerably improves the capability of direct visualization of the spatial distribution of defined bacterial populations inside the sludge flocs (Figures 4.3 and 5a). From a series of optical sections different types of image can be reconstructed. One option allows one to show all parts of for example an extended filamentous bacterium in focus (two-dimensional reconstruction; Figure 5a). Other options yield three-dimensional reconstructions such as red–green stereo images or coloured depth profiles (not shown). SCLM can also be used to analyse structural properties of activated sludge flocs such as floc size and homogeneity. M. Wagner, B. Assmus, R. Amann, P. Hutzler and A. Hartmann (submitted) demonstrated that most of the flocs of a municipal sewage treatment plant have been completely colonized by bacteria, whereas flocs containing gas bubbles or particles in the interior have rarely been detected.

4.5 Acknowledgements

This work was made possible by grants of the Deutsche Forschungsgemeinschaft (Am 73/2-2) and the Forschungsverbund FORBIOSICH to R.A. We thank H. Lemmer (Bayerische Landesanstalt für Wasserforschung, Munich, Germany) and D. Wedi (Lehrstuhl für Wassergüte- und Abfallwirtschaft, TU München, Munich, Germany) for helpful discussions and critically reading the manuscript. The help of Dr. P. Hutzler and Dr. B. Assmus (both GSF-Forschungszentrum für Umwelt und Gesundheit, Oberschleissheim, Germany) in SCLM is appreciated. We also thank Sibylle Schadhauser and Frank-Oliver Glöckner for excellent technical assistance.

4.6 References

Adamse, A.D. (1968) Formation and final composition of the bacterial flora of a dairy waste activated sludge. *Wat. Res.* **2**, 665–671.

Amann, R.I., Binder, B.J., Olson, R.J., Chisholm, S.W., Devereux, R. and Stahl, D.A. (1990a) Combination of 16S rRNA-targeted oligonucleotide probes with flow cytometry for analyzing mixed microbial populations. *Appl. Envir. Microbiol.* **56**, 1919–1925.

Amann, R.I., Krumholz, L. and Stahl, D.A. (1990b) Fluorescent oligonucleotide probing of whole cells for determinative, phylogenetic and environmental studies in microbiology. *J. Bact.* **172**, 762–770.

Amann, R.I., Springer, N., Ludwig, W., Görtz, H.-D. and Schleifer, K.-H. (1991) Identification *in situ* and phylogeny of uncultured bacterial endosymbionts. *Nature* **351**, 161–164.

Amann, R.I., Stromley, J., Devereux, R., Key, R. and Stahl, D.A. (1992) Molecular and microscopic identification of sulfate-reducing bacteria in multispecies biofilms. *Appl. Envir. Microbiol.* **58**, 614–623.

Auling, G., Pilz, F., Busse, H.-J., Karrasch, S., Streichan, M. and Schön, G. (1991) Analysis of the polyphosphate-accumulating microflora in phosphorus-eliminating, anaerobic–aerobic activated sludge systems by using diaminopropane as a biomarker for rapid estimation of *Acinetobacter* spp.. *Appl. Envir. Microbiol.* **57**, 3585–3592.

Benedict, R.G. and Carlson, D.A. (1971) Aerobic heterotrophic bacteria in activated sludge. *Wat. Res.* **5**, 1023–1030.

Blackall, L.L., Harbers, A.E., Greenfield, P.F. and Hayward, A.C. (1991) Foaming in activated sludge plants: a survey in Queensland, Australia and an evaluation of some control strategies. *Wat. Res.* **25**, 313–317.

Blackall, L.L., Parlett, J.H., Hayward, A.C., Minnikin, D.E., Greenfield, P.F. and Harbers, A.E. (1989) *Nocardia pinensis* sp. nov., an actinomycete found in activated sludge foams in Australia. *J. Gen. Microbiol.* **135**, 1547–1558.

Blackbeard, J.R., Ekama, G.A. and Marais, G.v.R. (1986) A survey of filamentous bulking and foaming in activated sludge plants in South Africa. *Wat. Pollut. Contr.* **85**, 90–100.

Bohlool, B.B. and Schmidt, E.L. (1980) The immunofluorescence approach in microbial ecology. *Adv. Microb. Ecol.* **4**, 203–241.

Brock, T.D. (1992) The genus *Leucothrix*, In *The prokaryotes* (ed. A. Balows, H.G. Trüper, M. Dworkin, W. Harder and K.-H. Schleifer), 2nd edn, pp. 3247–3255. Springer-Verlag, New York.

Brodisch, K.E.U. (1985) Interactions of different groups of microorganisms in biological phosphate removal. *Wat. Sci. Technol.* **17**, 89–97.

Buchan, L. (1983) Possible biological mechanism of phosphorus removal. *Wat. Sci. Tech.* **15**, 87–103.

Burggraf, S., Mayer, T., Amann, R., Schadhauser, S., Woese, C.R. and Stetter, K.O. (1997) Identifying *Archaea* with rRNA-targeted oligonucleotide probes. *Appl. Envir. Microbiol.*, in the press.

Cloete, T.E. and Steyn, P.L. (1988) A combined membrane filter-immunofluorescent technique for the in situ identification and enumeration of *Acinetobacter* in activated sludge. *Wat. Res.* **22**, 961–969.

DeLong, E.F., Wickham, G.S. and Pace, N.R. (1989) Phylogenetic stains: ribosomal RNA-based probes for the identification of single microbial cells. *Science* **243**, 1360–1363.

Devereux, R., Kane, M.D., Winfrey, J. and Stahl, D.A. (1992) Genus- and group-specific hybridization probes for determinative and environmental studies of sulfate-reducing bacteria. *Syst. Appl. Microbiol.* **15**, 601–609.

Dias, F.F. and Bhat, J.V. (1964) Microbial ecology of activated sludge. I. Dominant bacteria. *Appl. Microbiol.* **12**, 412–417.

Eikelboom, D.H. (1975) Filamentous organisms observed in activated sludge. *Wat. Res.* **9**, 365–388.

Embley, M.T. and Finlay, B.J. (1994) The use of small subunit rRNA sequences to unravel the relationships between anaerobic ciliates and their methanogen endosymbionts. *Microbiology* **140**, 225–235.

Emerson, D. and Ghiorse, W.C. (1992) Isolation, cultural maintenance, and taxonomy of a sheath-forming strain of *Leptothrix discophora* and characterization of manganese-oxidizing activity associated with the sheath. *Appl. Envir. Microbiol.* **58**, 4001–4010.

Farquhar, G.J. and Boyle, W.C. (1971) Occurence of filamentous microorganisms in activated sludge. *J. Wat. Pollut. Control Fed.* **43**, 779–798.

Foot, R.J., Kocianova, E. and Forster, C.F. (1992) Variable morphology of *Microthrix parvicella* in activated sludge systems. *Wat. Res.* **26**, 875–880.

Fuhs, G.W. and Chen, M. (1975) Microbiological basis of phosphate removal in the activated sludge process for the treatment of wastewater. *Microb. Ecol.* **2**, 119–138.

Gherna, R. and Woese, C.R. (1992) A partial phylogenetic analysis of the 'Flavobacter-Bacteroides' phylum: basis for taxonomic restructuring. *Syst. Appl. Microbiol.* **15**, 513–521.

Giovannoni, S.J., DeLong, E.F., Olsen, G.J. and Pace, N.R. (1988) Phylogenetic group-specific oligodeoxynucleotide probes for identification of single microbial cells. *J. Bact.* **170**, 720–726.

Goebel, U.B., Geiser, A. and Stanbridge, E.J. (1987) Oligonucleotide probes complementary to variable regions of ribosomal RNA discriminate between *Mycoplasma* species. *J. Gen. Microbiol.* **133**, 1969–1974.

Güde, H. (1980) Occurrence of cytophagas in sewage plants. *Appl. Envir. Microbiol.* **39**, 756–763.

Hernandez, M., Jenkins, D. and Beaman, B.L. (1993) Mass and viability estimations of *Nocardia* in activated sludge and anaerobic digesters using conventional stains and immunofluorescent methods. In *Abstracts of the First IAWQ International Specialized Conference on Microorganisms in Activated Sludge and Biofilm Processes*, 27–29 September 1993, Paris, pp. 245–255.

Hicks, R., Amann, R.I. and Stahl, D.A. (1992) Dual staining of natural bacterioplankton with 4,6-diamidino-2-phenylindole and fluorescent oligonucleotide probes targeting kingdom-level 16S rRNA sequences. *Appl. Envir. Microbiol.* **58**, 2158–2163.

Hiraishi, A. (1988) Respiratory quinone profiles as tools for identifying different bacterial populations in activated sludge. *J. Gen. Appl. Microbiol.* **34**, 39–56.

Hiraishi, A., Masumune, K. and Kitamura, H. (1989) Characterization of the bacterial population structure in an anaerobic–aerobic activated sludge system on the basis of respiratory quinone profiles. *Appl. Envir. Microbiol.* **55**, 897–901.

Howgrave-Graham, A.R. and Steyn, P.L. (1988) Application of the fluorescent-antibody technique for the detection of *Sphaerotilus natans* in activated sludge. *Appl. Envir. Microbiol.* **54**, 799–802.

Jenkins, D., Richard, M.G. and Daigger, G.J. (1986) *Manual on the causes and control of activated sludge bulking and foaming*. Ridgelines Press, Lafayette, CA.

Lane, D.J., Stahl, D.A., Olsen, G.J., Heller, D.J. and Pace, N.R. (1985) Phylogenetic analysis of the genera *Thiobacillus* and *Thiomicrospira* by 5S rRNA sequences. *J. Bact.* **163**, 75–81.

Larsen, N., Olsen, G.J., Maidak, B.L., McCaughey, M.J.,

Overbeek, R., Macke, T.J., Marsh, T.L. and Woese, C.R. (1993) The ribosomal database project. *Nucleic Acids Res.* **21**, 3021–3023.

Lawrence, J.R., Korber, D.R., Hoyle, B.D., Costerton, J.W. and Caldwell, D.E. (1991) Optical sectioning of microbial biofilms. *J. Bact.* **173**, 6558–6567.

Lemmer, H. (1986) The ecology of scum causing actinomycetes in sewage treatment plants. *Wat. Res.* **20**, 531–535.

Lemmer, H. and Baumann, M. (1988) Scum actinomycetes in sewage treatment plants. Part 3. Synergisms with other sludge bacteria. *Wat. Res.* **22**, 765–767.

Lötter, L.H. and Murphy, M. (1985) The identification of heterotrophic bacteria in an activated sludge plant with particular reference to polyphosphate accumulation. *Water SA*, **11**, 179–184.

Manz, W., Amann, R., Ludwig, W., Wagner, M. and Schleifer, K.-H. (1992) Phylogenetic oligodeoxynucleotide probes for the major subclasses of proteobacteria: problems and solutions. *Syst. Appl. Microbiol.* **15**, 593–600.

Manz, W., Wagner, M., Amann, R. and Schleifer, K.-H. (1994) In situ characterization of the microbial consortia active in two wastewater treatment plants. *Wat. Res.* **28**, 1715–1723.

Mulder, E.G. and Deinema, M.H. (1992) The sheathed bacteria. In: *The prokaryotes* (ed. A. Balows, H.G. Trüper, M. Dworkin, W. Harder and K.-H. Schleifer), 2nd edn, pp. 2612–2624. Springer-Verlag, New York.

Nakamura, K., Masuda, K. and Mikami, E. (1989) Polyphosphate accumulating bacteria and their ecological characteristics in activated sludge process. In: *Recent advances in microbial ecology* (ed. T. Hattori, Y. Ishida, Y. Maruyama, R. Morita and A. Uchida), pp. 427–431. Japan Scientific Societies Press, Tokyo, Japan.

Neefs, J.M., Vandepeer, Y., DeRijk, P., Chapelle, S. and DeWachter, R. (1993) Compilation of small ribosomal subnit RNA structures. *Nucleic Acids Res.* **21**, 3025–3049.

Nielsen, P.H. (1984) Oxidation of sulfide and thiosulfate and storage of sulfur granules in *Thiothrix* from activated sludge. *Wat. Sci. Tech.* **17**, 167–181.

Ohsumi, T., Shoda, M. and Udaka, S. (1980) Influence of cultural conditions on phosphate accumulation of *Arthrobacter globiformis* PAB-6. *Agric. Biol. Chem.* **44**, 325–331.

Olsen, G.J., Lane, D.J., Giovannoni, S.J., Pace, N.R. and Stahl, D.A. (1986) Microbial ecology and evolution: a ribosomal RNA approach. *Ann. Rev. Microbiol.* **40**, 337–365.

Palmer, C.J., Tsai, Y.-L., Paszko-Kolva, C., Mayer, C. and Sangermano, L.R. (1993) Detection of *Legionella* species in sewage and ocean water by polymerase chain reaction, direct fluorescent-antibody, and plate culture methods. *Appl. Envir. Microbiol.* **59**, 3618–3624.

Pike, E.B. and Curds, C.R. (1971) The microbial ecology of the activated sludge process. In *Microbial aspects of pollution* (*Soc. Appl. Bact. Symp. Ser.*, no. 1) (ed. G. Sykes and F.A. Skinner), pp. 123–147. Academic Press, London and New York.

Poulsen, L.K., Ballard, G. and Stahl, D.A. (1993) Use of rRNA fluorescence in situ hybridization for measuring the activity of single cells in young and established biofilms. *Appl. Envir. Microbiol.* **59**, 1354–1360.

Prakasam, T.B.S. and Dondero, N.C. (1967) Aerobic heterotrophic populations of sewage and activated sludge. I. Enumeration. *Appl. Envir. Microbiol.* **15**, 461–467.

Ramsing, N.B., Kühl, M. and Jorgensen, B.B. (1993) Distribution of sulfate-reducing bacteria, O_2, and H_2S in photosynthetic biofilms determined by oligonucleotide probes and microelectrodes. *Appl. Envir. Microbiol.* **59**, 3840–3849.

Reasoner, D.J. and Geldreich, E.E. (1985) A new medium for the enumeration and subculture of bacteria from potable water. *Appl. Envir. Microbiol.* **49**, 1–7.

Roller, C., Wagner, M., Amann, R., Ludwig, W. and Schleifer, K.-H. (1997) In situ probing of Gram-positive bacteria with high DNA G+C content using 23S rRNA-targeted oligonucleotides. *Microbiology*, in the press.

Rouf, M.A. and Stokes, J.L. (1964) Morphology, nutrition and physiology of *Sphaerotilus discophorus*. *Arch. Microbiol.* **49**, 132–149.

Saiki, R.K., Gelfand, D.H., Stoffel, S., Scharf, S.J., Higuchi, R., Horn, G.T., Mullis, K.B. and Ehrlich, H.A. (1988) Primer-directed enzymatic amplification of DNA with a thermostable DNA polymerase. *Science* **239**, 487–491.

Schaechter, M.O., Maaloe, O. and Kjeldgaard, N.O. (1958) Dependency on medium and temperature of cell size and chemical composition during balanced growth of *Salmonella typhimurium*. *J. Gen. Microbiol.* **19**, 592–606.

Seiler, H., Braatz, R. and Ohmayer, G. (1980) Numerical cluster analysis of the coryneform bacteria from activated sludge. *Zbl. Bakt. Hyg., I. Abt. Orig.* C**1**, 357–375.

Schleifer, K.H., Amann, R., Ludwig, W., Rothemund, C., Springer, N. and Dorn, S. (1992) Nucleic acid probes for the identification and *in situ* detection of pseudomonads. In *Pseudomonas: molecular biology and biotechnology* (ed. E. Galli, S. Silver and B. Witholt), pp. 127–134. American Society for Microbiology, Washington, DC.

Schmidt, T.M., DeLong, E.F. and Pace, N.R. (1991) Analysis of a marine picoplankton community by 16S rRNA gene cloning and sequencing. *J. Bact.* **173**, 4371–4378.

Soddell, J.A. and Seviour, R.J. (1990) Microbiology of foaming in activated sludge. *J. Appl. Bact.* **69**, 145–176.

Spring, S., Amann, R., Ludwig, W., Schleifer, K.-H. and Petersen, N. (1992) Phylogenetic diversity and identification of nonculturable magnetotactic bacteria. *Syst. Appl. Microbiol.* **15**, 116–122.

Stackebrandt, E. (1991) Unifying phylogeny and phenotypic diversity. In *The prokaryotes* (ed. A. Balows, H.G. Trüper, M. Dworkin, W. Harder and K.-H. Schleifer), 2nd edn, pp. 19–47. Springer-Verlag, New York.

Stackebrandt, E., Murray, R.G.E., Trüper, H.G. (1988) Proteobacteria classis nov., a name for the phylogenetic taxon that includes the 'purple bacteria and their relatives'. *Int. J. Syst. Bact.* **38**, 321–325.

Stahl, D.A. (1986) Unity in variety. Bio/Technology **4**, 623–628.

Stahl, D.A. and Amann, R. (1991) Development and application of nucleic acid probes in bacterial systematics. In *Sequencing and hybridization techniques in bacterial systematics* (ed. E. Stackebrandt and M. Goodfellow), pp. 205–248. John Wiley & Sons, Chichester.

Stahl, D.A., Flesher, B., Mansfield, H.R. and Montgomery, L. (1988) Use of phylogenetically based hybridization probes for studies of ruminal microbial ecology. *Appl. Envir. Microbiol.* **54**, 1079–1084.

Stahl, D.A., Lane, D.J., Olsen, G.J. and Pace, N.R. (1984) Analysis of hydrothermal vent-associated symbionts by ribosomal RNA sequences. *Science* **224**, 409–411.

Stahl, D.A., Lane, D.J., Olsen, G.J. and Pace, N.R. (1985) Characterization of a Yellowstone hot spring microbial community by 5S rRNA sequences. *Appl. Envir. Microbiol.* **49**, 1379–1384.

Streichan, M., Golecki, J.R. and Schön, G. (1990) Polyphosphate-accumulating bacteria from sewage plants with different processes for biological phosphorus removal. *FEMS Microb. Ecol.* **73**, 113–124.

Szwerinski, H., Gaiser, S. and Bardtke, D. (1985) Immunofluorescence for the quantitative determination of nitrifying bacteria: interference of the test in biofilm reactors. *Appl. Microbiol. Biotechnol.* **21**, 125–128.

Towner, K.J. (1992) The genus *Acinetobacter*. In *The prokaryotes* (ed. A. Balows, H.G. Trüper, M. Dworkin, W. Harder and K.-H. Schleifer), 2nd edn, pp. 3137–3143. Springer-Verlag, New York.

van Veen, W.L. (1973) Bacteriology of activated sludge, in particular the filamentous bacteria. *Antonie van Leeuwenhoek*, **39**, 189–205.

van Veen, W.L., van der Kooij, D., Geuze, E.C.W.A. and van der Vlies, A.W. (1973) Investigations of the sheathed bacterium *Haliscomenobacter hydrossis* gen.n., sp.n., isolated from activated sludge. *Antonie van Leeuwenhoek* **39**, 207–216.

Völsch, A., Nader, W.F., Geiss, H.K., Nebe, G. and Birr, C. (1990) Detection and analysis of two serotypes of ammonia-oxidizing bacteria in sewage plants by flow-cytometry. *Appl. Envir. Microbiol.* **56**, 2430–2435.

Wagner, M., Amann, R., Kämpfer, P., Assmus, B., Hartmann, A., Hutzler, P., Springer, N. and Schleifer, K.H. (1997) Identification and in situ detection of Gram-negative filamentous bacteria in activated sludge. *Syst. Appl. Microbiol.*, in the press.

Wagner, M., Amann, R., Lemmer, H. and Schleifer, K.-H. (1993) Probing activated sludge with oligonucleotides specific for proteobacteria: inadequacy of culture-dependent methods for describing microbial community structure. *Appl. Envir. Microbiol.* **59**, 1520–1525.

Wagner, M., Erhart, R., Manz, W., Amann, R., Lemmer, H., Wedi, D. and Schleifer, K.-H. (1994) Development of an rRNA-targeted oligonucleotide probe specific for the genus *Acinetobacter* and its application for in situ monitoring in activated sludge. *Appl. Envir. Microbiol.* **60**, 792–800.

Waitz, S. and Lackey, J.B. (1959) Morphological and biochemical studies on the organism *Sphaerotilus natans*. *Q. Jl Fl. Acad. Sci.* **21**, 335–340.

Wallner, G., Amann, R. and Beisker, W. (1993) Optimizing fluorescent in situ hybridization with rRNA-targeted oligonucleotide probes for flow cytometric identification of microorganisms. *Cytometry* **14**, 136–143.

Wanner, J. and Grau, P. (1989) Identification of filamentous microorganisms from activated sludge: a compromise between wishes, needs and possibilities. *Wat. Res.* **23**, 883–891.

Ward, D.M., Weller, R. and Bateson, M.M. (1990) 16S rRNA sequences reveal numerous uncultured microorganisms in a natural community. *Nature* **345**, 63–65.

Weller, R. and Ward, D.M. (1989) Selective recovery of 16S rRNA sequences from natural microbial communities in the form of cDNA. *Appl. Envir. Microbiol.* **55**, 1818–1822.

White, J.G., Amos, W.B. and Fordham, M. (1987) An evaluation of confocal versus conventional imaging of biological structures by fluorescence light microscopy. *J. Cell Biol.* **105**, 41–48.

Williams, T.M. and Unz, R.F. (1985) Filamentous sulfur bacteria of activated sludge: characterization of *Thiothrix*, *Beggiatoa*, and Eikelboom Type 021N strains. *Appl. Envir. Microbiol.* **49**, 887–898.

Woese, C.R. (1987) Bacterial evolution. *Microbiol. Rev.* **51**, 221–271.

Woese, C.R., Stackebrandt, E., Macke, T.J. and Fox, G.E. (1985a) A phylogenetic definition of the major eubacterial taxa. *System. Appl. Microbiol.* **6**, 143–151.

Woese, C.R., Weisburg, W.G., Hahn, C.M., Paster, B.J., Zablen, L.B., Lewis, B.J., Macke, T.J., Ludwig, W. and Stackebrandt, E. (1985b) The phylogeny of purple bacteria: the gamma subdivision. *System. Appl. Microbiol.* **6**, 25–33.

5. Principles in the modelling of biological wastewater treatment plants

M.C. Wentzel and G.A. Ekama

5.1 Introduction

Worldwide, an increasing awareness of the adverse impact that wastewaters have on aquatic environments (e.g. eutrophication) has led to the introduction of more stringent legislation controlling the water quality of effluents discharged from wastewater treatment plants. For example, legislation controlling discharges of the eutrophic nutrients nitrogen (N) and phosphorus (P) with municipal wastewater effluents are being widely applied (for example, in South Africa's 1980 amendment to Section 21 of the 1956 Water Act). To comply with the new effluent legislations, the functions of wastewater treatment plants have expanded from carbonaceous energy (measured via chemical oxygen demand (COD)) removal to include nitrification, denitrification and phosphorus removal, all mediated biologically. These extensions have been accomplished principally in activated sludge systems, through manipulation of the system configuration: incorporation of multiple in-series reactors, some aerated and others not, with various inter-reactor recycles. The principal objective of these manipulations is to create environmental conditions in the activated sludge system that are conducive to the growth of those organisms that naturally perform the biological reactions (termed **processes**) that are necessary to treat the wastewater. Consequently, not only have the system configuration and its operation increased in complexity, but concomitantly the number of biological processes influencing the effluent quality and the number of compounds involved in these processes have increased. With such complexity, designs based on experience will no longer give optimal performance; design procedures based on more fundamental behavioural patterns are required. Furthermore, with the increased complexity it is no longer possible to make reliable quantitative, or sometimes even qualitative, predictions as to the effluent quality to be expected from a design, or to assess the effect of a system or operational modification, without some model that predicts the system behaviour accurately.

This chapter describes some of the basic principles that are used in the development and validation of models for biological wastewater treatment systems. Modelling of the activated sludge system will be used as an example, as the authors are most familiar with this method for wastewater treatment.

5.2 Types of model

A model can be defined simply as a purposeful representation or description (often simplified) of a system of interest. In terms of this definition, models are widely used in science and engineering. For example, microbiologists and sociologists study model organisms (e.g. *Escherichia coli*) and model communities (Andrews 1993); engineers apply models in the design of a diverse variety of systems (e.g. wastewater treatment plants); researchers use scaled reproductions of full-sized objects for investigations (e.g. laboratory-scale experiments); and so on. Many different types of models exist; these can be broadly categorized into (1) physical, (2) verbal or conceptual and (3) mathematical. The physical model is a spatial scaled representation of the system. For example, the laboratory- and pilot-scale experiments used by scientists and engineers to investigate system response and behaviour are physical models. The verbal or conceptual model provides a qualitative description of the system and is usually developed from detailed observations; these models can be presented as schematic diagrams (e.g. flow diagrams) or as a series of narrative statements. The mathematical model provides a quantitative description of the system. With mathematical models the rates of the processes acting in the system and their stoichiometric interaction with the compounds are formulated mathematically. The mathematical formulations need to be incorporated in a solution procedure that takes account of the physical constraints and characteristics imposed by the system in which the processes take place, e.g. temperature and mixing conditions.

For the design and operation of biological wastewater treatment systems, it is the mathematical models that have proved most useful. By providing quantitative descriptions, they allow predictions of the system response and performance to be made. From the predictions design and operational criteria can be identified for optimization of system performance. Also, mathematical models can serve as very powerful research tools. By evaluating model predictions it is possible to test hypotheses on the behaviour of the wastewater treatment system (e.g. biological processes, their response to system constraints) in a consistent and integrated fashion. This may direct attention to issues not obvious from the physical system, and lead to a deeper understanding of the fundamental behavioural patterns controlling the system response. In essence, mathematical models can provide a defined framework that can direct thinking (design, operation or research). Although this is usually very advantageous, it does have some disadvantages: the framework can restrict innovative and new developments.

This chapter will focus on the development of mathematical models. However, mathematical models are seldom developed in isolation, but usually evolve interactively from a conceptual model that might be based to some degree on observations made on a physical model (e.g. laboratory- or pilot-scale experiments). Accordingly, these aspects will also receive some attention.

5.3 Mathematical model objectives

The objective of a mathematical model is to give quantitative expression to behavioural patterns of interest in a system. In wastewater treatment, the various uses of mathematical models can be summarized as follows. A model:

- gives expression to conceptual ideas, to account for major events of interest occurring within a system, and allows evaluation of these; by comparing the simulated and the observed response, attention can be drawn to deficiencies in the conceptual structure
- provides information not apparent from pilot-scale studies; this can be particularly useful in the system being modelled is a complex one
- allows potentially feasible solutions to be explored that are not covered by pilot-scale and other studies, thereby giving guidance for selecting the more promising ones for testing
- assists in identifying the parameters that significantly influence the system response and thereby gives guidance for the establishment of design criteria
- assists in identifying possible causes for system malfunction or failure, and in devising remedial measures.

It is unlikely that any particular mathematical model will fulfil all of these uses. Simpler models can be developed to satisfy specific uses; these will have restricted use but also usually require less input than the more complex models, with their greater range. A mathematical model can be deemed successful if it fulfils the expectations of it.

5.4 Mathematical models and levels of organization

Two extremes in mathematical models can be identified: empirical and mechanistic. An empirical model is based on recognition of the parameters that seem to be essential to describe the behavioural pattern of interest, and linking these by empirical relationships established by observation; the mechanisms and/or processes operating in the system are not known or are ignored. This is the classical 'black box' approach. In contrast, a mechanistic model is based on some conceptualization of the biological/physical mechanisms operating in the system, i.e. it is based on some conceptual model. The complexity of this conceptual model will depend on the degree of understanding of the biological, physical and chemical processes occurring in the system. Because the mechanistic models have some conceptual basis, they often are more reliable than the empirically based models: because of their 'black box' approach, the empirical models have application strictly limited to remain in the boundaries (e.g. wastewater characteristics, system parameters) within which the model was developed; only interpolation is possible. Being conceptually based, the mechanistic models have greater sureness in application outside the boundaries within which the model was developed; both interpolation and extrapolation are possible. However, eventually all models are only our rationalization of behavioural patterns of parameters we conceive to be of interest. Owing to this rationalization, any model needs to be rigorously calibrated and adequately verified by appropriate tests. Also, the conditions within which the model is expected to operate successfully need to be firmly delineated: for the empirical models these are strictly the conditions within which the model was developed; for the mechanistic models these are the conditions under which the conceptualized behaviour is expected to remain valid. It is evident from the discussion above that the mechanistically based models have greater potential for application to wastewater treatment plants, and attention will be focused on these models.

To set up the conceptual model on which the mechanistic mathematical model is based, the processes operating in the system and the compounds on which these act are identified, and the various interactions between the processes, and between the processes and compounds, are delineated descriptively. From the conceptual model, the process rates and their stoichiometric interactions with the compounds are formulated mathematically, to develop the mechanistic model. The conceptual model, and its mathematical equivalent the mechanistic model, very probably will not include all the processes and compounds that are present in the system; only those conceived to be of significance for fulfilling the objectives set for the model need be included. The art of constructing the conceptual and mechanistic models is in eliminating those processes and compounds that contribute little or nothing to fulfilling the objectives set for the model: it is a waste of time and effort to develop a complicated model where a simpler one is adequate. Indeed, it is most unlikely that a model can be developed that describes a phenomenon completely; theoretically a complete description should include aspects down to the most fundamental level. To take an extreme example, in the description of a biosystem, for completeness DNA molecule behaviour should be included, clearly an impossible task. Instead less complete models are developed at some level of organization (Odum 1971); only characteristics important at that level are incorporated in the model description. The level of organization is usually set by the objectives for the model. For example, in modelling biological behaviour in wastewater treatment systems, we cannot directly implicate biochemical control mechanisms (such as ADP/ATP and NAD/NADH ratios), or even the behaviour of specific microorganism species: the mixed liquor in the activated sludge system contains a wide diversity of different microorganism species for which identification and enumeration techniques are only now becoming available. Instead, microorganisms that fulfil a particular function in the activated sludge system (e.g. aerobic degradation of organics or nitrification) are grouped together as a single entity, which has been called a 'surrogate' organism. This surrogate organism is assigned a set of unique characteristics that reflect the behaviour of the group but might not reflect the characteristics of any individual organism or species of organisms in the group. To illustrate, this approach is equivalent to modelling the 'macroscopic' behaviour of a forest of trees as opposed to the 'microscopic' behaviour of each individual tree or species of trees that make up the forest. In considering the behaviour of the forest, a parameter that could be modelled, for example, is carbon dioxide (CO_2) production. The forest as an entity will have defined specific CO_2 production and consumption rates; individual tree species within the forest, or even every individual tree, might have specific CO_2 production and consumption rates that deviate significantly from those of the forest entity. However, the effect achieved by modelling the forest as an entity will closely equal the net effect of modelling the cumulative contribution of each individual tree or tree species. The great advantage in modelling the forest as an entity over modelling the individuals is that considerably less information is required to develop the model and to calibrate it. In modelling biological wastewater treatment systems, the utilization of substrate by organisms is a typical example: Monod's equation (Monod 1949) is used to relate the specific growth rate of the surrogate organism to the surrounding substrate concentration, whereas the organisms making up the surrogate group might have different specific growth rates or might respond differently to the various substrates present in the influent wastewater.

Thus, for modelling wastewater treatment systems the organizational level that is modelled is the mass behaviour of a population or group of selected microorganisms. In the models developed for activated sludge systems, the principle organism groups, their functions and the zones in which these functions are performed are summarized in Table 5.1.

Accepting the level of organization detailed above, the parameters at that level that need to be included in the mathematical model depend greatly on the objectives for the model. For mathematical modelling of wastewater treatment systems, two different kinds of mathematical models are generally developed, steady-state and dynamic. The steady-state models have constant flows and loads and tend to be relatively simple. This simplicity makes these models useful for design; in these models complete descriptions of system parameters are not required, but rather the models are oriented towards determining the more important system design parameters. The dynamic models have varying flows and loads and accordingly include time as a parameter; the dynamic models are more complex than the steady-state ones. The dynamic models are useful in predicting time-dependent system response of an existing or proposed system; their complexity, however, means that for application the system parameters have to be completely defined. For this reason the use of dynamic models for design is restricted. Often the steady-state design and dynamic kinetic models evolve interactively: The dynamic kinetic models can provide guidance for the development of the steady-state design models; they help identify the design parameters that have a major influence on the system response and help eliminate those processes that are not of major importance at steady state.

For the dynamic models, with their greater complexity, only those parameters that seem to

Table 5.1. Principal organism groups included in models for activated sludge systems, their functions and the zones in which these functions are performed

Organism	Biological process	Zone
1. Heterotrophs		
1.1 Obligate anaerobes	COD removal (organic degradation, DO uptake)	Aerobic
	Ammonification (organic N → NH_4^+)	Aerobic
1.2 Facultative	As for 1.1	Aerobic
	Denitrification (organic degradation, $NO_3^- \rightarrow NO_2^- \rightarrow N_2$	Anoxic
	Fermentation (F-RBCOD → SCFA)	Anaerobic
2. Poly-P heterotrophs	P release (SCFA uptake, PHA storage)	Anaerobic
	P release (SCFA uptake, PHA storage)	Anoxic
	P uptake (PHA degradation, denitrification?	Anoxic
	P uptake; P removal (PHA degradation, DO uptake)	Aerobic
3. Chemical autotrophs	Nitrification ($NH_4^+ \rightarrow NO_2^- \rightarrow NO_3^-$, DO uptake)	Aerobic

be of importance are considered for inclusion in the model. For activated sludge systems, selecting the level of organization as the surrogate organism or mass behaviour of populations, until recently the dynamic models have been structured to consider only the net effects as present in the bulk liquid. For example, in using Monod's equation the kinetic rate has been determined by the bulk liquid soluble COD and surrogate organism concentrations. However, with the extentions of the models to include biological excess phosphorus removal (BEPR), parameters internal to the surrogate biomass have had to be included, e.g. polyhydroxyalkanoate (PHA) and polyphosphate (Wentzel *et al.* 1992). With this development, although the model might be at a selected level of organization, information on processes and behaviour from lower levels of organization is often essential, particularly to identify the key processes that control the response of the system. Usually information from lower levels of organization is microbiological and/or biochemical, and the more complete this information is the more reliable the model. To make use of this information, 'model' organisms that are part of the 'surrogate' are identified and the known microbiological and biochemical characteristics of the organism are used to obtain a greater understanding of the surrogate. For example, BEPR can be accomplished by a number of organisms but *Acinetobacter* spp. has been used as a 'model' organism and its known characteristics have been used to gain an insight into the processes mediated by the surrogate organism group responsible for BEPR.

More recently the surrogate organism approach to modelling has been found to be inadequate to describe completely some behavioural patterns observed in activated sludge systems; for example the selector effect (Gabb *et al.* 1991), substrate utilization inhibition on transfer from anoxic to aerobic zones (Casey *et al.* 1993), and generation of nitrogen intermediates in denitrification (Casey *et al.* 1993; Wild *et al.* 1994). To describe these and similar observations, it has been found that a lower level of organization needs to be selected: the synthesis and activity of certain key enzymes and the processes they mediate need to be modelled (Wild *et al.* 1994; Kappeler *et al.* 1994). Modelling at this level of organization has been termed **modelling with structured biomass**. Detailed microbiological and biochemical information is required for this modelling approach.

5.5 Modelling tasks

From the above, it is apparent that to develop a mechanistically based mathematical model for wastewater treatment systems, a number of tasks need to be completed:

- Identify objectives for the model.
- Describe the conditions within which the model is to operate.
- Identify the essential compounds utilized and formed.
- Identify the processes acting on these compounds.
- Conceptualize a mechanistic model that qualitatively describes the kinetic and stoichiometric behaviour of the processes and compounds.
- Formulate mathematically the process rates and stoichiometry.
- Set up a solution procedure that incorporates the process rates, stoichiometry and transport terms.
- Calibrate the model and test its response against that observed experimentally.

Each of these tasks is discussed briefly below. In the discussion it will become apparent that these tasks cannot be completed sequentially, but rather the model tends to evolve interactively with the tasks being undertaken. For example, to identify processes and compounds, one needs to have some initial conceptualization of the system behaviour, i.e. a rudimentary conceptual model. As more information becomes available from model application and testing, aspects of the rudimentary model are improved as new compounds and processes are identified for inclusion, or processes already included are modified.

5.5.1 Objectives of the model

The final use for the model needs to be identified; if it is for design, steady-state models are usually adequate, if for simulation a dynamic model is required. The steady-state models are very much simpler than the dynamic in that wherever possible stoichiometric relationships are used and kinetics are excluded. Owing to the inclusion of time as a parameter in the dynamic models, they require inclusion of kinetic relationships. Also, the objectives of the model should be matched to the functions that are required of the biological treatment system. For example, the additional complexity of biological P removal need not be included in the model if this is not a required function of the treatment system.

5.5.2 Model conditions

In this task, the restrictions under which the model will operate must be identified. For example, a model might be restricted to activated sludge systems treating municipal wastewaters principally of domestic origin in a temperature range of 14–22 °C.

5.5.3 Compounds

Compounds directly involved in the processes of interest need to be identified. The selection of compounds is influenced by the level of organization of the model, and the extent of understanding of the biological, physical or chemical processes that act on the various compounds. Some of the compounds are directly measurable (e.g. ammonium, orthophosphate, oxygen) whereas the means for measuring other compounds are not available, or where they are available they are not sensitive enough to give quantitative parameters. Accordingly, for these latter compounds their existence has to be inferred, either from hypothesized behaviour or from the requirement of mass balances and continuity principles. Examples of inferred compounds in wastewater treatment models are organism active mass, endogenous mass and inert mass which together make up the measurable parameter volatile suspended solids (VSS).

5.5.4 Processes

The processes (biological, physical or chemical reactions) that act on the compounds above are identified by observing changes in the compounds under a variety of conditions, or by applying mass balance and continuity principles. Identification is largely influenced by the objectives set for the model (i.e. functions of the wastewater treatment plant), the level of organization selected for the modelling (molecular, species, population, etc.) and understanding of the mechanisms operative at the selected level of organization. Also, information that is available at the lower levels of organization can play a crucial role in identifying important processes. For example, microbiological and biochemical information has helped identify the key processes that control the phenomenon of biological excess phosphorus removal.

5.5.5 Conceptual model

The conceptual model is developed interactively with identification of the compounds and processes and is a verbal description of the compounds and processes and their interactions. It is also usually based on understanding the chemistry, microbiology, biochemistry, etc., of the mechanisms that control the processes. However, the conceptual model takes a more macroscopic approach than the often detailed chemical, microbiological and biochemical information available. For example, a more microscopic approach would require modelling each biochemical reaction and the mechanisms controlling it, such as ATP/ADP or $NADH/NAD^+$ ratios; there is usually no quantitative experimental data for structuring such a model, and the complexity introduced with such detail is not warranted.

5.5.6 Mathematical formulation

The process rates and their stoichiometric interactions with the compounds are formulated mathematically. These formulations are usually based on extensive experimental observations and the requirement for mass balances (i.e. conservation of COD, N, P, etc). Basing formulations on experimental data means that these often have a large empirical component. For modelling of growth processes, Monod functions are generally used relating 'surrogate' organism specific growth rate to substrate concentration. For surrogate organism death, the endogenous respiration approach of Herbert (1958) or the death–regeneration approach of Dold *et al.* (1980) is used; the Pirt (1975) maintenance energy concept is not used as it cannot explain organism death in the absence of an external substrate source or electron acceptor, conditions commonly encountered in biological wastewater treatment systems.

5.5.7 Solution procedure

The process rates and stoichiometric interactions with the compounds define the behaviour at a single point in the system. To obtain the response of the system, the following need to be incorporated: system configuration (single or multiple reactors), reactor type (continuously fed, batch, etc), hydraulic mixing regime in the reactor (plug flow or completely mixed), solids regime (suspended or fixed), recycle flow between reactors, and mass transport of compounds in and out of each reactor. Methods for incorporating these parameters are described in the next section.

5.5.8 Calibration and testing

Biological wastewater treatment models contain a number of kinetic and stoichiometric constants that need to be quantified. Great care must be taken in this task; often the constants are interactive so that infinite sets of constants can be derived that give the same predictions. For example, Marais and co-workers (Marais and Ekama 1976; Dold *et al.* 1980; Van Haandel *et al.* 1981) found that description of the behaviour of activated sludge systems must include the growth (substrate utilization) and endogenous mass loss processes. In the stoichiometry of growth, for every unit of substrate consumed, it is accepted that a constant fraction (the true specific yield) appears as new cell mass and the remainder is oxidized to generate energy for synthesis, giving rise to an associated oxygen requirement. When attempting to obtain an estimate of the true specific yield, Marais and co-workers found that it cannot be determined directly from sludge production in systems operated under steady state because the active mass synthesized is subject to endogenous/maintenance/death processes. These processes not only reduce the active mass but also generate inert (endogenous) residues, both particulate and soluble, so that the observed volatile solids not only are less than that indicated by the true specific yield, but are also made up of active and endogenous particulate fractions, to give an apparent yield. Owing to these endogenous effects, the apparent yield decreases as the sludge age of the system increases. If it is accepted that the endogenous residue generation is constant with respect to the active mass that disappears, it is possible to obtain estimates of the true specific yield and endogenous mass loss rates by trial and error curve-fitting to observed yields in a set of steady-state responses over a range of sludge ages. However, Marais and Ekama (1976) found that the yield and endogenous mass loss effect act in a compensatory fashion; good 'fits' can be obtained by different pairs of true specific yield values and endogenous mass loss rates. They found it essential to obtain an independent assessment of one of the processes. To achieve this, they established that the endogenous mass loss rate could be isolated by doing aerobic digestion batch tests on samples of mixed liquor without addition of substrate. From a semi-logarithmic plot of the oxygen utilization rate against time, they determined that endogenous mass loss conformed to first-order kinetics with respect to the active mass and that the slope of the plot defined the specific endogenous mass loss rate. Applying this rate to the steady-state systems over a range of sludge ages, and accepting the value for the unbiodegradable particulate endogenous residue fraction of the active mass measured by McCarthy and Brodersen (1962) and Washington and Hetling (1965), they determined the true specific yield.

Taking due cognizance of the above, essentially there are three ways whereby the constants can be quantified: from a test in which the effect of the constant is isolated and directly measured; from a test in which the effect of the constant is completely dominant compared with the effect of other constants; and by 'curve fitting', using a range of operating conditions – this approach can be applied only if most of the other constants have been evaluated. Generality of the constants can be greatly improved if values can be obtained from a variety of sources, including information available at levels of organization lower than that being modelled. For example, principles in bioenergetics can be used to confirm specific organism yields; pure culture studies of specific species can provide information on growth rates, and so on.

With the constants determined, the model can gain credibility and acceptance only through widespread application to simulate successfully the responses observed over a wide range of system configurations, operational modes and compositions of influent wastewaters. Credibility is enhanced if this application is successful without the kinetic and stoichiometric constants being changed. The predictions of the model are only as good as the data on which the calibration is based.

5.6 Model presentation

In the mathematical model, the process rates and the stoichiometric relationships between the processes and compounds are formulated mathematically. With the steady-state design models, their presentation does not present undue difficulty because of their simplicity. However, with the dynamic models the large number of complex interactions between compounds and processes necessitates that these be clearly presented. The IAWQ Task Group (Henze *et al.* 1987) on 'Mathematical Model-

Table 5.2. Monod–Herbert process model in matrix format

Compound $i \rightarrow$ j Process ↓	1 Heterotrophic biomass, Z_{BH} (M (COD)/l^3)	2 Soluble substrate, S_{bs} (M (COD)/l^3)	3 Compound oxygen, O (M (–COD)/l^3)	Process rate, ρ_j
1 Growth	1	$-1/Y_{ZH}$	$-(1 - Y_{ZH})/Y_{ZH}$	$\mu_H \frac{S_{bs}}{(K_{SH} + S_{bs})} Z_{BH}$
2 Decay	–1		–1	$b_H Z_{BH}$
Observed conversion rates (M l^{-3} t^{-1})		$r_i = \sum_j \nu_{ij} \rho_j$		

ling of Wastewater Treatment' have recommended the matrix method for dynamic model presentation. This format facilitates clear and unambiguous presentation of the compounds and processes and their interaction on a single page. In addition, the matrix format allows easy comparison of different models, and facilitates transforming the model into a computer program. The setting up of such a matrix, how to interpret it and how it is incorporated in the mathematical solution procedures are described briefly below, using a simple model as an example.

5.6.1 Setting up the matrix (process model)

Table 5.2 presents, in matrix format, the essential components of a simple Monod–Herbert process model for aerobic microbial growth on a soluble substrate, accompanied by endogenous mass loss (for simplicity, endogenous residue generation is specifically excluded).

The matrix is represented by a number of columns and rows; one column for each compound and one row for each process. The first step in setting up the matrix is to identify the compounds of relevance in the model. The Monod–Herbert model quantifies the growth of the heterotrophic biomass compound (Z_{BH}, COD units) at the expense of the soluble substrate compound (S_{bs}, COD units). By keeping track of Z_{BH} and S_{bs}, it is possible to calculate the oxygen requirement; in this fashion oxygen (O) can be included as a third compound. (Note that oxygen has units of negative COD because oxygen is not a product of this process: electrons are passed to oxygen to form water, but the generation of water cannot be measured and so *consumption* of oxygen, which can be readily measured, is modelled instead.) The compounds are presented as symbols listed at the head of the appropriate column, and are defined (with dimensions) at the bottom of the corresponding columns. The index i is assigned to identify a compound in the range of compounds. In this case, i ranges from 1 to 3 for the three compounds considered in this simple model.

The second step in setting up a matrix is to identify the biological processes occurring in the system. These are conversions or transformations that affect the compounds considered in the model. Only two processes take place in this simple model: aerobic growth of organisms at the expense of soluble substrate, and endogenous mass loss. These are itemized one below the other down the left-hand side of the matrix. The index j is assigned to identify the process in the range of processes. In this case, j can assume a value of 1 or 2.

The process rates are formulated mathematically and are listed down the right-hand side of the matrix in line with the respective process. These process rates are given the symbol ρ_j, with j denoting the index of the biological process.

Along each process row the stoichiometric coefficient for conversion from one compound to another is inserted so that each compound column lists the stoichiometric coefficients for the processes that influence that compound. The stoichiometric coefficients are given the symbol ν_{ij}, where i denotes the index of the compound and j the index of the process. In our example, for process 1 (j = 1, aerobic growth of heterotrophs) the compound heterotrophic biomass (Z_{BH}) increases (+1), the compound soluble substrate (S_{bs}) decreases ($-1/Y_{ZH}$) and the compound oxygen (O) decreases [$-(1 - Y_{ZH})/Y_{ZH}$].

The sign convention used in the matrix for each compound *per se* is 'negative for consumption' and 'positive for production'. Cognizance must be taken of the units used in the rate equations for the processes. For example, the rate equation for the process aerobic growth of heterotroph biomass, ρ_1, is written as a biomass growth rate (not as a substrate utilization rate) and has units of mg of cell COD growth/mg of substrate COD utilized per day. The stoichiometric values are thus normalized with respect to the heterotroph biomass concentration (Z_{BH}), i.e. the stoichiometric coefficients for Z_{BH} and

S_{bs} are 1 and $-1/Y_{ZH}$ respectively, not Y_{ZH} and -1.

The stoichiometric coefficients, v_{ij}, are greatly simplified by working in consistent units; in the example above, the compounds are expressed as COD equivalents. Provided that consistent units are used, continuity can be checked from the stoichiometric parameters by moving across any row of the matrix; with consistent units, the sum of the stoichiometric coefficients must be zero (noting that oxygen is equivalent to negative COD).

This matrix forms a succinct summary of the complex interactions between compounds and processes. The matrix in effect constitutes a fingerprint uniquely characterizing the mathematical model. It allows alterations in processes, compounds, stoichiometry and kinetics to be readily incorporated. The matrix representation method has two main benefits:

- It allows the effect of a particular process on the compounds to be easily determined, as follows: the reader moves along a particular row, i.e. process, and multiplies the stoichiometric coefficient (v_{ij}) by the process rate (ρ_j). This gives the reaction rate (r_{ij}) for the particular compound being affected by the single process, i.e.
$$r_i = v_{ij}\rho_j. \tag{5.1}$$
In representing the matrix, by adding up the reaction rates for a particular process, a mass balance must be obtained.
- It allows rapid and easy recognition of the fate of each compound, as follows: the reader moves down the column representing the compound of interest and multiplies the stoichiometric coefficient (v_{ij}) by the process rate (ρ_j). The summation of these multiplications gives the overall reaction rate (r_i) for the compound, i.e.
$$r_i = \sum_j v_{ij}\rho_j. \tag{5.2}$$

5.6.2 Switching functions

Under certain conditions the process rate equations are not operative, e.g. aerobic processes are not operative under anaerobic conditions. For example, consider the aerobic growth of biomass; in Table 7.2 the Monod growth rate equation (Monod 1949) has been used:

$$\rho_1 = \frac{\mu_H S_{bs}}{(K_{SH} + S_{bs})} Z_{BH}. \tag{5.3}$$

In an environment where the dissolved oxygen concentration (O) is zero (or perhaps close to zero), the rate of this aerobic process should also be zero. Mathematically, this can be achieved by multiplying the Monod rate expression by a 'switching' factor that is zero when O is zero, and unity when the environment is 'fully' aerobic. Experience has shown that a switching function formulation that is very flexible and useful is one that takes the same form as the Monod formulation. For example, the switching function for oxygen is:

$$\frac{O}{(K_O + O)}, \tag{5.4}$$

where K_O is a switching constant of small magnitude (say 0.1 mg O/l).

The selection of a small value for K_O means that the value of the switching function decreases from near unity to zero at very low O values, i.e. when the DO value decreases below about 0.2 mg O/l. However, the function is mathematically continuous, which helps to eliminate problems of numerical instability in simulating system behaviour; such problems can arise if the rate is switched 'on' and 'off' discontinuously.

On incorporating the switching function the process rate equation then becomes:

$$\rho_1 = \frac{\mu_H S_{bs}}{(K_{SH} + S_{bs})} \frac{O}{(K_O + O)} Z_{BH}. \tag{5.5}$$

With the switching function operating on the Monod growth rate equation, when O is zero the value of the switching function is zero, and the process rate, ρ_1, will be zero. However, if O is say 1 mg O/l then the value of the switching function is close to unity (i.e. 'fully' aerobic) and the process rate will be given by the Monod growth rate equation. In this way, the process of aerobic growth is switched 'on' or 'off' automatically by the model depending on the dissolved oxygen concentration.

In certain situations, the switching 'off' of one process might be linked to the switching 'on' of another. If, for example, the oxygen input to a nitrifying activated sludge system were terminated periodically, there would be a switch from aerobic to anoxic growth. The latter process is governed by kinetic and stoichiometric expressions that differ from those for the aerobic growth process. To account for this phenomenon in a single model, the rate equations for aerobic and anoxic growth can be multiplied by the appropriate switching functions as follows:

$$\rho_{\text{aerobic}} = \rho_{\text{aerobic}} \frac{O}{(K_O + O)}; \tag{5.6}$$

$$\begin{aligned} \rho_{\text{anoxic}} &= \rho_{\text{anoxic}} \left(1 - \frac{O}{(K_O + O)}\right) \\ &= \rho_{\text{anoxic}} \frac{K_O}{(K_O + O)}. \end{aligned} \tag{5.7}$$

In this instance, it is apparent that the selection of K_O will influence the point at which there is a switch from aerobic to anoxic growth, and vice versa. That is, K_O now influences the model predictions and is not only serving a mathematical objective. Therefore, whenever switching functions are used, care should be taken in the selection of the magnitude of the switching constant (K_O here) to ensure that the model predictions are not incorrectly biased.

5.6.3 Matrix solution

Solution of the matrix can be fixed in time (e.g. batch test), space (e.g. steady-state multiple reactor system), or time and space (e.g. multiple reactor system with time-varying flow).

5.6.3.1 Solution in time

This solution requires that the initial concentration be known, whereafter changes in concentration are determined by integrating forward in time. Integration forward follows the basic Euler equation or equivalent:

$$C(t + \Delta t) = C(t) + \left(\frac{dC}{dt}\right)_t \Delta t, \qquad (5.8)$$

where

C = compound concentration
t = time
Δ = step size in integration
$(dC/dt)_t$ = reaction rate.

The reaction rate is obtained from the summation down the particular compounds column of the multiplication terms $\nu_{ij}\rho_j$, as described previously.

5.6.3.2 Solution in space

Solution of the matrix in space requires that the transport terms be included (generally hydraulic terms for wastewater treatment systems e.g. flows in and out of the reactor). Inclusion of the transport terms and the rate equations is facilitated by use of the mass balance equation:

$$\begin{pmatrix}\text{mass rate}\\ \text{of}\\ \text{accumulation}\end{pmatrix} = \begin{pmatrix}\text{mass rate}\\ \text{of}\\ \text{input}\end{pmatrix} - \begin{pmatrix}\text{mass rate}\\ \text{of}\\ \text{output}\end{pmatrix} + \begin{pmatrix}\text{mass rate of}\\ \text{production}\\ \text{by reaction}\end{pmatrix}. \qquad (5.9)$$

The mass of input and output are the transport terms and depend on the physical characteristics of the system being modelled. The mass of production for a particular compound is obtained from the matrix. Taking an example, in symbols, for a completely mixed reactor:

$$\frac{V\,dC_{out}}{dt} = Q_{in}C_{in} - Q_{out}C_{out} - r_i V, \qquad (5.10)$$

where

V = volume
Q_{in} = flow rate in
Q_{out} = flow rate out
C_{in} = concentration of compound in influent flow
C_{out} = concentration of compound in outflow (i.e. reactor concentration for completely mixed reactor)
r_i = reaction rate obtain from the matrix; see equation (5.2)
dC_{out}/dt = rate of change of reactor concentration of compound C.

Dividing by V and recognizing that at steady state $dC/dt = 0$,

$$\frac{Q_{in}}{V}C_{in} - \frac{Q_{out}}{V}C_{out} - r_i = 0. \qquad (5.11)$$

Mass balance equations are derived for each compound in every reactor (including the settler). This yields a set of simultaneous nonlinear equations for each reactor, which can then be solved to give values for all the compounds. As the equations are nonlinear, repetitive techniques must be employed in the solution (Billing 1987; Billing and Dold 1988a,b,c).

5.6.3.3 Solution in time and space

This solution is confined to the situation of single or multiple reactors under repetitive diurnal flows. Again mass balances are set up, but unlike the steady-state system the transients (dC/dt) no longer equal zero. Initial concentration values are selected and the mass balance equations are integrated forwards until the solution is reached; this is achieved when the concentrations of all compounds in each reactor at the start and at the end of the diurnal cycle are equal (Billing 1987; Billing and Dold 1988a,b,c).

For example, from Table 5.2, the rate of reaction for the compound $i = 1$, heterotrophic biomass (Z_{BH}) at a point in the system would be:

$$r_1 = \frac{\mu_H S_{bs}}{(K_{SH} + S_{bs})} Z_{BH} - b_H Z_{BH}. \qquad (5.12)$$

Similarly for the compound $i = 2$, soluble substrate (S_{bs}):

$$r_2 = \frac{1}{Y_{ZH}} \frac{\mu_H S_{bs}}{(K_{SH} + S_{bs})} Z_{BH} \qquad (5.13)$$

and for the compound $i = 3$, dissolved oxygen (O):

$$r_3 = \frac{1 - Y_{ZH}}{Y_{ZH}} \frac{\mu_H S_{bs}}{(K_{SH} + S_{bs})} Z_{BH} - b_H Z_{BH}. \qquad (5.14)$$

To create the mass balance for any compound within a given system boundary (e.g. a completely mixed reactor), the conversion rate, r_i, would be multiplied by the reactor volume and added to the appropriate advective terms, i.e. input and output masses (where each mass is given by concentration multiplied by flow in or out of the system) for the particular system; this mass balance is not shown here as the system is not defined in this presentation.

The rate of production by reaction, r_i, might

be of interest on its own. For example, equation (5.14) defines the 'rate of production' of O; therefore $-r_3$ defines the oxygen utilization rate at a point within the system. This parameter is often of interest in aerobic systems.

5.7 Conclusion

In this chapter some of the basic principles for the development of models for biological wastewater treatment systems have been described. Models can serve as extremely useful tools in the design and operation of wastewater treatment plants, and in research into the behaviour of these plants. For design, models can provide guidance in identifying the key design parameters and can quantify system parameters to ensure optimal performance. For operation, models can provide quantitative predictions as to the effluent quality to be expected from a design or existing system, and allow the effect of system or operational modifications to be assessed theoretically. For research, models allow testing of hypotheses in a consistent and integrated fashion, to direct attention to issues not obvious from the physical system and so lead to a deeper understanding of the fundamental behavioural patterns controlling the system response. In this manner models provide a defined framework that can guide the direction for further enquiry. However, this framework does have disadvantages; it can restrict innovative new developments that do not fall within boundaries of the framework. Also, in modelling and using models, it should be remembered that models are only our rationalization of behavioural patterns of parameters that we conceive to be of importance when describing a particular system. Owing to this rationalization, the models need to be adequately verified by appropriate tests and the conditions within which the model is expected to operate successfully need to be clearly defined. A model can be deemed successful if it fulfils the expectations of it.

5.8 References

Andrews, J.F. (1993) Modeling and simulation of wastewater treatment processes. *Wat. Sci. Tech.* **28** (11/12), 141–150.

Billing, A.E. (1987) Modelling techniques for biological systems. M.Sc. thesis, Dept. Chem. Eng., Univ. of Cape Town, Rondebosch 7700, South Africa.

Billing, A.E. and Dold, P.L. (1988a) Modelling techniques for biological reaction systems. 1. Mathematic description and model representation. *Wat. SA* **14** (4), 185–192.

Billing, A.E. and Dold, P.L. (1988b) Modelling techniques for biological reaction systems. 2. Modelling of the steady state case. *Wat. SA* **14** (4), 193–206.

Billing, A.E. and Dold, P.L. (1988c) Modelling techniques for biological reaction systems. 3. Modelling of the dynamic case. *Wat. SA* **14** (4), 207–218.

Casey, T.G., Ekama, G.A., Wentzel, M.C. and Marais, G.v.R. (1993) An hypothesis for the causes and control of low F/M filamentous organism bulking in nitrogen (N) and nutrient (N & P) removal activated sludge systems. In *Proc. of the IAWQ First Int. Conf. on Microorganisms in Activated Sludge and Biofilm Processes*, Paris, 27–28 September.

Dold, P.L., Ekama, G.A. and Marais, G.v.R. (1980) A general model for the activated sludge process. *Prog. Wat. Tech.* **12**, 47–77.

Gabb, G.M.D., Still, D.A., Ekama, G.A., Jenkins, D. and Marais, G.v.R. (1991) The selector effect on filamentous bulking in long sludge age activated sludge systems. *Wat. Sci. Tech.* **23** (Tokyo), 867–877.

Henze, M., Grady, C.P.L. Jr, Guyer, W., Marais, G.v.R. and Matsuo, T. (1987) *Activated sludge model No. 1* (IAWQ Scientific and Technical Report No. 1). IAWQ, London.

Herbert, D (1958) Recent progress in microbiology. In *VII International Congress for Microbiology* (ed. G. Tunevall), pp. 381–396. Almquist and Wiksel, Stockholm.

Kappeler, J., Casey, T.G., Wentzel, M.C., Ekama, G.A. and Marais, G.v.R. (1994) Activated sludge modelling with structured biomass. *Wat. Res.* (submitted).

Marais, G.v.R. and Ekama, G.A. (1976) The activated sludge process: Part I. Steady state behaviour. *Wat SA* **2** (4), 163–200.

McCarthy, P.L. and Broderson, C.F. (1962) Theory of extended aeration activated sludge. *J. Wat. Pollut. Control Fed.* **34**, 1095–1103.

Monod, J. (1949) *The growth of bacterial cultures*. Pasteur Institute, Paris.

Odum, E.P. (1971) *Fundamentals of ecology*. WB Saunders Co., London.

Pirt, S.J. (1975) *Principles of microbe and cell cultivation*. Blackwell Scientific Publications, Oxford.

Van Haandel, A.P.C., Ekama, G.A. and Marais, G.v.R. (1981) The activated sludge process. 3. Single sludge denitrification. *Wat. Res.* **15**, 1135–1152.

Washington, D.R. and Hetling, L.J. (1965) Volatile sludge accumulation in activated sludge plants. *J. Wat. Pollut. Control Fed.* **37**, 499–507.

Wentzel, M.C., Ekama, G.A. and Marais, G.v.R. (1992) Processes and modelling of nitrification dentrification biological excess phosphorus removal systems. *Wat. Sci. Tech.* **25** (6), 59–82.

Wild, D., Schulthess, R.V. and Guyer, W. (1994) Synthesis of denitrification enzymes in activated sludge: modelling with structured biomass. Presented at 17th IAWQ Biennial Conference, Budapest, July.

6. Practical considerations for the design of biological wastewater treatment systems

WISA-Group

6.1 Basic design data

6.1.1 General introduction

This chapter is concerned only with domestic sewage from residential and commercial areas and is for sewage works treating up to 5 Ml/d of domestic sewage.

The design of any new sewage treatment works, or the extension to an existing works, should be based on the following information:

(1) volumetric assessment of the flow rate, which will determine all hydraulic aspects of the works
(2) sewage strength according to various parameters, which will determine the biological and chemical processes in the works.

If an existing works has to be extended and records of flow, sewage constituents and plant performance are available, the future trends could be based on these data. A careful study of the rate of flow, the fluctuation in sewage strength and the performance data of the various existing units in the plant will be of considerable value.

If, however, no reliable data on the existing or the proposed new works are available, then estimates have to be made. The main purpose of this chapter is to provide basic guidelines in this respect.

6.1.2 Evaluation of available information

6.1.2.1 Flow recordings

The reliability of recorded flow figures should be verified. Simple checks should be made such as comparing flow charts over an extended period, observing service records of flow-measuring instruments and comparing calculated flow rates at the measuring flume with flow records.

Determine the following flow figures from the records:

(1) the total flow during dry weather over 24 h on seven consecutive days will give the average dry weather flow, i.e. ADWF in kl/day
(2) the peak rate of flow over short periods in dry weather will give the peak dry-weather flow, i.e. PDWF in l/s
(3) the peak rate of flow over short spells during rainy conditions with high storm-water inflow will give the peak wet-weather flow, i.e. PWWF in l/s.

Rainfall recordings, conditions of sewer systems and enforcement of by-laws regarding storm-water ingress into sewers could be taken into account when the above-mentioned flow figures are evaluated.

6.1.2.2 Raw sewage analysis

If sample analyses are available it is important to know the sampling method and the time interval between flow measurement and sampling. Composite samples taken over 24 h are more reliable than single samples. It is preferable, however, to base the calculation of loads on analyses of synchronized proportional sampling and flow recordings.

The following parameters will enable the quality of the raw sewage to be evaluated:

- settleable solids (Set Sols – 1 h settlement)
- suspended solids (SS)
- oxygen absorbed in 4 h (OA)
- biochemical oxygen demand (BOD)
- chemical oxygen demand (COD)
- ammonia (NH_4^+-N)
- total Kjeldahl nitrogen (TKN)
- phosphate (P)
- pH.

6.1.2.3 Analysis of plant performance

It is necessary to inspect the functioning of an existing works to establish possible shortcomings in its design and to assess the maximum capacity of the various units. Apart from more usual defects such as incorrect grit removal, inadequate operational facilities and excessive corrosion, samples taken at strategic positions will indicate the performance of each specific process stage. The quality of the final effluent should be aligned with the standard requirements. The interpretation of these analyses should be performed by persons experienced in this type of work.

Consultation with the plant operator and the person in charge on the history and the performance of the works should be seen as essential background for evaluating plant performance.

Table 6.1. Typical figures for fully sewered housing units with bathrooms, basins and kitchen sinks

Housing standard	No. of houses per net hectare	Litres/ha per day
Below average	20 or more	80–150
Average	3–20	150–220
Above average	3 or less	220–500

6.1.2.4 Design assumptions

For the final design parameters, it is important to assess flow and load figures as set out later in this chapter, and the information derived from the above-mentioned evaluations should be seen as supplementary. Should the theoretical and practical data differ widely, further investigation will be necessary.

6.1.3 Assessment of flow

Infiltration of groundwater and ingress of surface stormwater into the sewerage system will affect rate of flow significantly. Hence flow recording during both wet and dry weather is essential.

6.1.3.1 Domestic sewage

This component is normally the major contributor to the total flow; a reliable estimate is therefore important. There are various ways in which this may be determined:

(1) Domestic sewage flow may be expressed in terms of l/head per day. Typical figures for fully sewered housing units with bathrooms, basins and kitchen sinks are shown in Table 6.1. Flows in excess of those shown are likely where water supplies are not metered, and lower flows will prevail in residential areas without normal internal plumbing in the dwellings. No reliable correlation between the sewage flow and the number of sanitary fittings in a dwelling has yet been established.

(2) Water consumption records, where available, are a possible alternative basis for estimating domestic sewage flows. About 70% of domestic water consumption can be expected to reach the sewer, but this percentage will be higher from dwellings where there are no gardens, and lower where there are large gardens, especially if lawn sprinklers are used.

(3) It has been established in certain areas that a figure of 1 kl/day per stand provides a reasonable estimate of flow.

(4) Flows from other units contributing domestic sewage are estimated in Table 6.2.

In providing for the full design life of a sewage works the upward trend in per capita sewage contribution should be kept in mind and an annual increase of 1.5–2.5% could be expected.

6.1.3.2 Average dry weather flow (ADWF)

The flow used when designing treatment units will be the total domestic sewage flow and is known as the ADWF; it is measured in kl/d.

6.1.3.3 Peak dry weather flow (PDWF)

The normal daily peak flow, without allowing for stormwater inflow, exceeds the AWDF by a factor that is generally highest when the population and the area served are small (Table 6.3). The Harmon formula is often used to calculate the factor:

$$\text{Peak factor} = 1 + \frac{14}{4 + \sqrt{p}}$$

where p is the equivalent population (thousands) contributing to the system; e.g. $p = 2$ for 2,000 people. PDWF is measured in l/s.

The effect of the attenuation in long sewers is to decrease the peak factor. A sewered area with an elongated shape and flat slopes will likewise tend to lower the peak factor.

The peak dry weather flow for a works receiving sewage by pumping may vary considerably from that of a fully gravitational system. Such cases should be investigated in more detail.

6.1.3.4 Groundwater and stormwater

The average dry weather flow (ADWF) should exclude any provision for groundwater infiltration or stormwater entering the sewerage system. Capacity should, however, be provided for this peak wet weather flow. Special circumstances, such as high rainfall intensity, high natural water table, water tightness of pipe systems and frequency of drainage inspections, should be taken into account in assessing the peak wet weather flow.

6.1.3.5 Peak wet weather flow (PWWF)

Estimates of peak wet weather flow (PWWF) vary so widely that no criteria can be given. One empirical method for estimating increased flow under wet weather conditions is to allow for a flow rate equal to one and a half times the peak dry weather flow (PDWF).

6.1.4 Assessment of sewage strength

The strength of sewage arriving at treatment works varies considerably, depending largely on the domestic living standards of the contributing population.

Sewage characteristics can be divided into three main categories:

(1) the concentration of oxidizable organic material, or **substrate**

(2) the concentration of nutrients present

(3) the solids concentration.

Table 6.2. Sewage flows from other units contributing domestic sewage

Contributor	Unit	Sewage flow (litres/day)
People residing within the drainage area:		
School hostels	Per bed	100–200
Hotels	Per bed	200–400
Hospitals	Per bed	400–800
Tourist camps, motels and caravan parks	Per capita	100–150
People visiting from place outside the drainage area:		
Day workers (shops, offices)	Per capita	70
Factory workers	Per capita	120
Day schools, primary	Per capita	20
Day schools, secondary	Per capita	40
Restaurants	Per seat	30
Swimming pools	Per user	50
Drive-in theatres	Per bay	15

6.1.4.1 Organic concentration

The following are the most common parameters for measuring organic strength:

(1) chemical oxygen demand (COD)
(2) the five-day biochemical oxygen demand (BOD)
(3) oxygen absorbed (OA).

The selection of one of these parameters depends largely on practical consideration and on personal preferences. However, the following guidelines are presented:

(4) The OA is used in the design of biological filters; it does, however, provide a quick indication of the strength of a sample
(5) The BOD is used in the design of lagoon and pond systems, of biological filters and the activated sludge process. However, the test itself has a number of disadvantages making the result dependent on the reaction rate as well as the substrate concentration
(6) The COD test is used increasingly in the design of all biological processes. The test itself takes about 4 h and the procedure is complex.

In domestic sewages there is a fairly constant COD/BOD ratio of about 2:1 and in COD/OA of about 10:1. Thus, treatment processes for domestic sewage can be designed with COD, BOD or OA as a design parameter.

6.1.4.2 Nutrient concentration

The two most important nutrients in sewage treatment processes are nitrogen and phosphorus.

Nitrogen concentration is generally determined as total Kjeldahl nitrogen (TKN), which measures the sum of the free and saline ammonias and the organic nitrogen concentrations.

In most domestic sewage, phosphorus is present in both the polyphosphate and the orthophosphate forms. It is therefore important, when measuring the total phosphorus concentration present, to select a test that converts all the phosphate into one form, usually the orthophosphate.

Table 6.3. Peak factor according to size of populations served

Population served	Peak factor
2,000	3.5
10,000	3.0
30,000	2.5

6.1.4.3 Typical domestic sewage

The daily per capita load can be expressed by various parameters and these will also vary depending on the diet and social structure of the population served. Table 6.4 suggests loads for design purposes.

On the basis of the above-mentioned assumptions of flow and load per capita, typical domestic sewage will have the following general characteristics:

- OA (as O_2) 70–80 mg/l
- BOD (as O_2) 350–400 mg/l
- COD (as O2) 700–800 mg/l
- Settleable solids 10–12 ml/l
- Suspended solids 250–350 mg/l
- Ammonia NH_4^+-N) 40–45 mg/l
- Phosphate (P) 10–13 mg/l

If the major contribution to the sewage flow originates from lower-income groups, an increase in the BOD/COD concentrations should be expected.

6.1.4.4 Implementation

The design parameters arrived at from the preceding estimates should be moderated by reference to appropriate expertise and consultation with suitable qualified persons.

Table 6.4. Suggested design loads for a typical domestic sewage

Parameter	Design loads (g/head per day)
Oxygen demand:	
OA (as O_2)	10–16
BOD (as O_2)	50–80
COD (as O_2)	100–160
Solids:	
Settleable solids	60
Suspended solids	30
Nutrients:	
Ammonia nitrogen (NH_4^+-N)	15
Phosphate (P)	4

6.2 Design parameters

6.2.1 Design parameters for biological filtration

6.2.1.1 Load assessment

The load applied to a filter is the rate at which substrate or nutrient is applied to the biomass on the media and can be expressed as the product of strength and volume (see §6.1). This is expressed as g COD/OA/m^3 per day based on ADWF.

6.2.1.2 Design loadings

Table 6.5 is a guide to loading rates for biological filters to provide an acceptable quality effluent, based on the following conditions:

(1) winter conditions implying one month in which the monthly average sewage temperature is 15 °C
(2) medium is crushed rock of 40–63 mm size
(3) analysis is of composite sample
(4) the OA of a composite sample of settled sewage taken at regular intervals through the day does not exceed 75 mg/l.

Higher temperature, smaller size stone (within limits) and lower strength sewage are all factors that might permit higher loadings to be used. However, the higher the quality sought in the final effluent, the less this latitude in loading becomes.

It should be noted that wide variations in the performance of biological filters at similar organic loading rates occur and that the values cited in Table 6.5 are indicative.

Loading rates for biological filters in areas not subjected to severe winter conditions, where for example the minimum monthly average sewage temperature is 18 °C or above, can be increased by between 25% and 40% of the figures given in Table 6.5.

6.2.1.3 Dosing siphon and tank

The dosing siphon should be designed for the PWWF plus a safety factor. The dosing siphon tank should be sized to ensure that the maximum retention period will not exceed 1 h at ADWF.

6.2.1.4 Distributors

The distributor should be designed for the PWWF plus a safety factor.

6.2.1.5 Practical considerations

(i) Granular filter media

The choice of medium is generally limited by what is available locally. For maximum treatment per volume of filter, hard strong media with a high specific surface should be selected provided that the void ratio is not unacceptably reduced.

Specific surface areas and void ratios have not been measured in the past, so performance in relation to these parameters is not readily available. The current parameters, which are empirical, are the surface characteristics of the medium and its size.

It is thought that for a given size and a given loading, a medium with an irregular and rough surface will produce a better effluent than one with a comparatively smooth regular surface.

Stone media, packed directly above the filter aeration tiles, shall be within the grading −15 mm to +100 mm.

The following nominal sizes are recommended in works where no special strength problems arise:

- Single-stage filtration — 50 or 63 mm
- Double-stage filtration
  - Primary filter — 50 or 63 mm
  - Secondary filter — 40 mm

Uniformity of media has been stressed in the past. However, some small range of grading might be permissible in giving additional media surface as shown in Table 6.6, based on British Standard Specification 1438:1971, *Media for biological percolating filters*, which should be the criterion.

Media should be substantially free of fine material, particularly clays, and organic matter.

(ii) Plastic media

A number of proprietary makes of plastic medium are on the market, the characteristics of which should be ascertained from the manufacturers. These generally have a very high void ratio and therefore have less tendency to pond when strong wastes are treated on them. The economics of the use of materials should be carefully considered. Experience has indicated that effluent treated on plastic filter media will not generally nitrify.

Table 6.5. Design loading according to filter depth and mode of operation

Filter depth and mode of operation	Design loading (g COD/m^3 per day) to provide effluent quality of: <75 mg/l COD	<10 mg/l N/NH$_4$ average	<10 mg/l N/NH$_4$ 90 percentile
2 m single	300	160	80
2 m single with recirculation	250	200	100
2 m double	250	200	100
4 m single	250	200	100
3 m single with recirculation	300	250	120
4 m double	350	300	150

(iii) Filter aeration
An important feature in the design of biological filters is the provision for adequate aeration. This can be achieved by covering the filter floor with suitable aeration tiles, with provision for air to enter the tiles from a central channel, inlet chambers or through holes in the filter wall. Vent pipes inside the filter in lieu of holes in the wall are undesirable.

Stone media, packed directly above the filter aeration tiles, shall be within the grading -150 mm to +100 mm. This layer shall be man-handled and handpacked very carefully around the aeration tiles without breaking, cracking or disturbing the tiles. This handpacked layer shall continue until at least 250 mm cover is reached above the highest points of the tiles.

(iv) Distributors
The distributor should be so designed as to distribute sewage evenly over the filter surface. Particular care should be taken to ensure that the distributor rotates at all times when sewage is being applied. A dosing siphon should be used to achieve this.

Reaction-driven distributors frequently operate with a rotational speed of 40–60 s per revolution, but slower speeds in the range 75–300 s per revolution have been found in practice not to have materially affected effluent quality.

The materials used for the centre column, distributor arms, etc., should be durable and non-corrosive.

(v) Depth of filter
Biological filter plants have the advantage of being able to operate without the use of electric or diesel power. Thus the design of the works should be based on gravity flow; this might determine the depth of filter, with 2 m depth being the minimum. However, deeper filters are more economic, as will be seen in the table of design loading rates (Table 6.5).

(vi) Dosing siphon tanks
Dosing siphon tanks accommodate the siphon, which will generally be supplied with the distributor. The siphon should be of a durable non-corrosive material.

Table 6.6. Media for biological percolating filters adapted from British Standard specification

BS 410 Square Hole Perforated Plate test sieves	Nominal size (% by mass passing) 63 mm	50 mm	40 mm
Passing 75 mm	100	—	—
Passing 63 mm	85–100	100	—
Passing 50 mm	0–35	85–100	100
Passing 37.5 mm	0–5	0–30	85–100
Passing 28 mm	—	0–5	0–40
Passing 20 mm	—	—	0–5

In cases where the initial daily flow is only a fraction of the design ADWF, the retention time in the dosing tanks can exceed the recommended 1 h. In such a case it would be necessary to provide a temporary wall within the tank to reduce the tank volume.

(vii) Potential corrosion
A high degree of nitrification in effluent with a low alkalinity might be the cause of severe corrosion. This can be overcome by either providing for denitrification by recirculation back to an anoxic zone, or by adding lime to raise the alkalinity. Microbiological induced corrosion (MIC) by sulphate-reducing bacteria (SRB) has been observed in centre columns and distributer arms with poor ventilation; for this reason it is essential that the materials of construction are protected against corrosion by means of suitable paint/resin linings.

6.2.2 Design parameters for rotating biological contactor (RBC)

6.2.2.1 Introduction

Process performance below design expectations, structural problems with shafts and media, excessive biomass buildup on media, hydraulic limitation and other process problems have been serious concerns at many installations in South Africa. These problems have resulted in the process's falling out of favour with designers in recent years. Although the process might still be used in some situations, its limitations

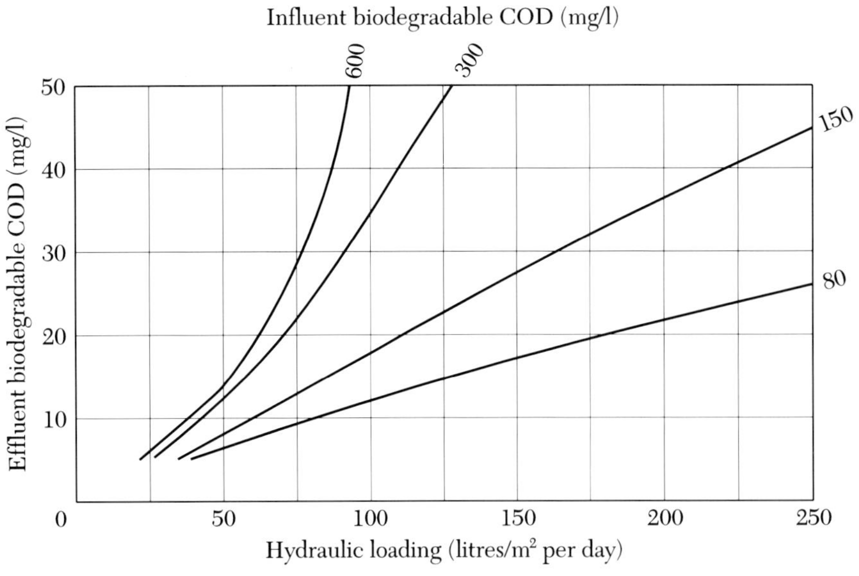

Figure 6.1. Design relationship for COD removal from domestic sewage, for temperatures above 14 °C.

Table 6.7. Correction factor for reducing the required wetted area when employing staging

Number of stages	Correction factor
3	0.95
4	0.90
More than 4	0.86

and potential problems must be understood and allowed for if it is to be used successfully.

6.2.2.2 Wetted surface area required

A family of curves showing the effect of the hydraulic loading on the effluent substrate concentration (biodegradable COD) for different influent substrate concentrations is shown in Figure 6.1. As a first approximation these curves should be used to determine the RBC's wetted surface area required for a particular septic tank effluent.

6.2.2.3 Effect of staging

The arranging of discs in a series of stages has been shown to increase treatment efficiency significantly. Different microbial biomass develops in each successive stage adapted to the changed composition resulting from successive stages of treatment. Carbonaceous oxidation of biodegradable COD occurs at the initial stages, followed by nitrification in the latter stages.

The RBC process exhibits, within limits, first-order kinetics. The improved residence time distribution obtained with staging increases the substrate removal rate. When more than two stages are provided the required wetted area can be slightly reduced, as shown in Table 6.7.

When staging is used, the organic load on the initial stages can become very high. To prevent diminishing performance and the occurrence of odours, an organic loading of not more than 100 g biodegradable COD/m^2 per day is considered an upper limit.

6.2.2.4 Nitrification

Nitrification will occur only when the biodegradable COD has been decreased to about 25 mg/l or less. Under those conditions and provided that the alkalinity is sufficient to maintain a pH greater than 7, nitrification will be influenced by the hydraulic loading, as shown in Figure 6.2.

6.2.2.5 Effect of population, size and flow variation

The design relationships given above have been determined for steady-state conditions. If a predetermined constant quality effluent is required (i.e. General Standard), provision should be made either to balance the flow and the load or a correction factor should be used for sizing the RBC. Suggested correction factors used for domestic sewage are given in Table 6.8.

Because of the balancing effect of septic tanks these factors should be used with circumspection on properly designed septic tank/RBC systems.

6.2.2.6 Effect of temperature

The temperature of sewage has little effect on RBC performance between 14 and 30 °C. A correction factor for cold areas could be used, as shown in Table 6.9.

6.2.2.7 Effect of rotational speed

The rotational speed is one of the few variables that the designer can control. When this is increased, the percentage substrate removal increases to a limiting value determined by the particular RBC under consideration. It has been reported that the percentage substrate removal increases with the rotational speed raised to the 0.1 power, whereas the corresponding power required to rotate the contactors increases at an exponential rate. As it is difficult

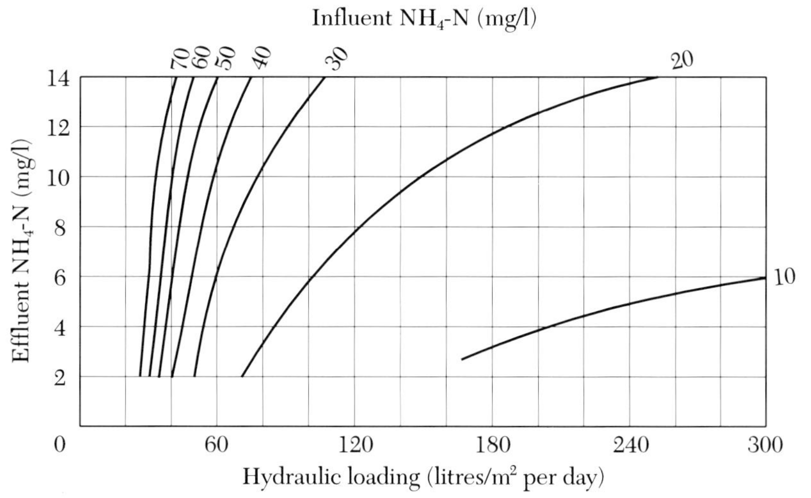

Figure 6.2. Design relationship for nitrification based on influent ammonium ion concentration, for temperatures above 14 °C.

to find a common basis for comparing the effect of different rotational speeds on contactors of different diameter, almost all RBC process studies have attempted to maintain speeds of between 1 and 5 rev./min, with little improvement noted at higher speeds. An economic evaluation should be made between increased power costs, and increased contactor area.

6.2.2.8 Empirical design formula

The design parameters given above will enable an efficient design to be achieved. As a design check the following formula should ensure that a septic tank/RBC system treating domestic sewage will produce an effluent with an OA concentration of 10 mg/l at a temperature of 15 °C:

Disc area (measured both sides) ≥5 m^2 per cap.

6.2.2.9 Final sedimentation tank

The biomass produced by RBCs is usually in large flocs and has relatively high settling velocities.

6.2.2.10 Recycling

Denitrification is often required to ensure complete nitrification. With a septic tank/RBC system this can be achieved by recycling, as shown schematically in Figure 6.3.

6.2.2.11 Practical considerations

(1) A simple inlet works should be provided, although grit removal is not essential with septic tanks.

(2) Extended power failures might cause drying of exposed sections of the discs, resulting in an imbalance. On subsequent start-up this might cause excessive torque resulting in gear stripping. Mechanical and electrical equipment should be sized accordingly.

(3) RBCs are relatively capital-intensive and are therefore more suited to situations of steady sewage flow.

Table 6.8. Correction factors for domestic sewage

Population	Correction factor
400	1.3
400–1,500	1.3–1.1
1,500–5,000	1.1–1.0

Table 6.9. Correction factors for cold areas

Temperature (°C)	Correction factor
14	1.00
12	1.05
10	1.15
8	1.30
6	1.40

(4) For practical design aspects of septic tanks refer to *A guide to the use of septic tank systems in South Africa* (CSIR Publication K 86-1985).

6.2.3 Design parameters for activated sludge

6.2.3.1 Sludge age and solids loading rate

This is the primary parameter in determining plant design as it affects all other variables. Activated sludge plants in South Africa operate at large sludge ages (at least 15 days), at which nitrification is generally assured and a relatively stable sludge generated. Loading rates are inversely related to sludge age. Permissible loading may thus be assessed in terms of kg COD/day per kg MLSS, in accordance with the values given in Table 6.10.

6.2.3.2 Activated sludge mass

The total mass required depends on the sludge age (or solids loading rate), the minimum reactor temperature and the biodegradable fraction of the incoming sewage.

This mass of active sludge comprises not only that contained in the (aerobic) reactor, but also

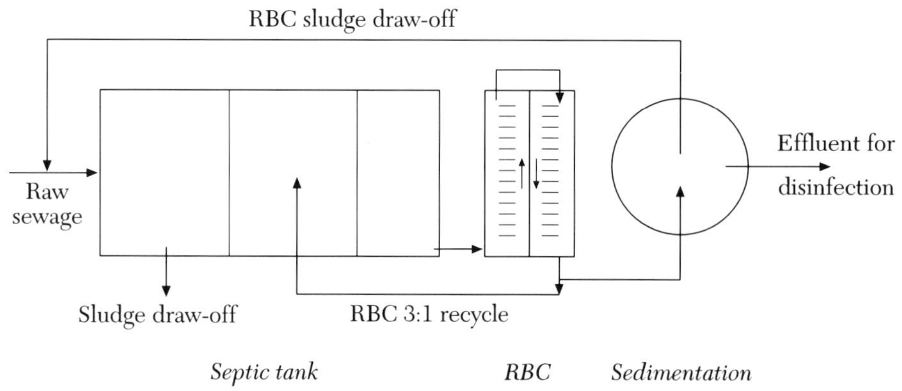

Figure 6.3. Schematic flow diagram showing effluent recycle and sludge return.

that in the final clarifier and the sludge return system. Design procedures result in a more conservative design, if the mass of the active sludge in the reactor only is taken into account.

Sludge age will determine the yield factor for new biomass produced per unit of COD loading, while at the same time limiting the loading rate as explained in §6.2.3.1 above. The approximate relationships between these three parameters are given in Table 6.10.

The sludge mass in the system will then be equal to the product of the sludge age, the yield factor and loading in kg COD per day.

6.2.3.3 Reactor volume

The reactor volume is not a primary design parameter but is determined by the required sludge mass divided by the design MLSS.

Thus if the total sludge mass is M_t kg and the total mixed liquor solids concentration is X_t kg/m^2, the reactor volume (V_R) can be expressed as:

$$V_R = M_t/X_t \text{ m}^3.$$

Equally the reactor volume can be expressed in terms of volatile solids, i.e.

$$V_R = M_v/X_v \text{ m}^3.$$

6.2.3.4 Waste activated sludge

The volume of sludge to be wasted each day (m^3/d) is dependent on sludge age and is calculated as follows:

(a) when wasting directly from the reactor:

$$\frac{\text{volume of reactor}}{\text{sludge age}}.$$

(b) when wasting from the returned activated sludge stream:

$$\frac{\text{volume of reactor} \times \text{MLSS}}{\text{sludge age} \times \text{RSS}},$$

where RSS is the concentration of solids in the returned sludge.

6.2.3.5 Nitrification

The relationship between the sludge age required for nitrification and the temperature is given by:

$$R = 3.05 \times (1.127)^{20-T},$$

where R is the minimum sludge age required for nitrification and T is the temperature, in degrees Celsius, of the mixed liquor in the reactor.

6.2.3.6 Return activated sludge

If p is the ratio of returned sludge to wastewater flow, RSS the concentration of the solids in the returned sludge and MLSS the mixed liquor concentration in the reactor then:

$$\text{RSS} = \frac{(1+p)}{p}\text{MSS}.$$

Values of p in excess of 1.5 are seldom encountered. Thus if returned sludge concentration falls, the flow of return sludge would need to be increased, although this would provide a diminishing return, as indicated by the above equation.

6.2.3.7 Oxygen requirements

The carbonaceous oxygenation requirement is a function of the strength of the sewage, measured as COD, BOD or OA and sludge age (or sludge loading rate). Empirical values that can be used as an alternative to other more mathematical approaches in use are given in Table 6.11.

The oxygen demand for nitrification is directly proportional to the mass of nitrogenous compounds to benitrified, the relationships being approximately 4.5 mg O_2/mg ammonia as N converted.

6.2.3.8 Capacity of aeration equipment

There are several different types of aeration equipment available. The transfer efficiency of the different devices differs in accordance with such parameters as power intensity and reactor geometry but typical figures for each type are given in Table 6.12. These figures represent the transfer efficiency into clean water of low TDS at 20 °C and at sea level, assuming zero residual DO and that the water contains no surfactants or other ingredients that could influence the rate of oxygen transfer.

Table 6.10. Relationships between sludge loading rate, sludge growth index and recommended sludge density

Sludge age (days)	Sludge loading rate (kg COD applied/day per kg MLSS)	Sludge production (kg MLSS produced/kg COD applied)		Recommended sludge density (mg/l)
		Settled sewage	Raw sewage	
30	0.154	0.22	0.23	5,000
25	0.174	0.23	0.25	4,400
20	0.200	0.25	0.27	3,900
15	0.242	0.28	0.31	3,500
10	0.312	0.32	0.36	3,300
5	0.506	0.40	0.45	3,200

Table 6.11. Oxygen requirement relationships to sludge loading rate and sludge age

Sludge loading rate (kg COD applied/day per kg MLSS)	Sludge age (days)	Dissolved oxygen required for carbonaceous oxidation (kg O_2/kg COD removed)	Ratio of max. to mean oxygenation rate	Ratio of min. to mean oxygenation rate
0.514	30	0.84	1.45	0.5
0.714	25	0.82	1.47	0.5
0.200	20	0.79	1.50	0.5
0.242	15	0.73	1.55	0.5
0.312	10	0.65	1.67	0.5
0.516	5	0.53	1.80	0.5

The transfer efficiency of aerators operating in sewage will be significantly lower than the above figures owing to the presence of surfactants and various other impurities. It is therefore necessary to apply an appropriate factor to the specified 'oxygenation capacity' to arrive at 'oxygen input'. This factor is known as the α factor and is generally taken to be 0.85.

Oxygen input is also influenced by altitude, temperature, residual DO and the positioning of aerators in the basin.

A further correction, known as the β factor, must therefore be applied. The combined effects of these factors are illustrated in Figure 6.4.

6.2.3.9 Clarifier design

If the settling characteristics of the mixed liquor are known, preferably in terms of the stirred specific volume index (SSVI) (in ml/g), a better clarifier design will result if the settling flux theory is used. In the absence of the above settling characteristics, solids flux loadings should not exceed 6 kg SS/m^2 per hour at peak flows.

6.2.3.10 Practical considerations

(i) Conventional designs

(a) *Plant layout*. The activated sludge process can be used as the complete treatment stage when preceded only by an inlet works comprising bar screens and grit channels. It can also be used as the secondary stage of treatment after primary sedimentation.

The conventional arrangement is an aeration tank followed by a clarifier with a returned activated sludge system from the clarifier to the inlet of the aeration tank.

Table 6.12. Oxygen transfer efficiency for different types of aeration

Type of aeration	Transfer efficiency (kg O_2/kW h)
Fine bubble diffused aeration	2.5–3.5
Low-speed surface aeration	1.8–2.6
Submerged turbine aerators	1.7–2.3
High-speed surface aeration	1.2–1.6
Coarse bubble aeration	1.4–1.8

Note: In this table the power is shaft and not motor power.

(b) *Aeration tank*. For small works a square or rectangular aeration tank fitted with one surface aerator for the former and two surface aerators for the latter might be adequate. It should be noted that both greater flexibility and better operation are achieved with a larger number of small aerators.

Aeration can also be achieved by means of a bubble aeration system, but whereas the efficiency can be greater with a fine bubble system, maintenance requirements are more exacting, so this is not recommended for small works.

Care must be taken to avoid the entrapment and accumulation of scum at any point in the system. This can lead to excessive growths of filamentous organisms from time to time, which are both unsightly and deleterious to the efficient operation of the process.

(c) *Clarifiers*. Circular clarifiers are usually recommended, with central baffled inlet, peripheral overflow weir and a continuously operating sludge scraper mechanism. A siphon scraper

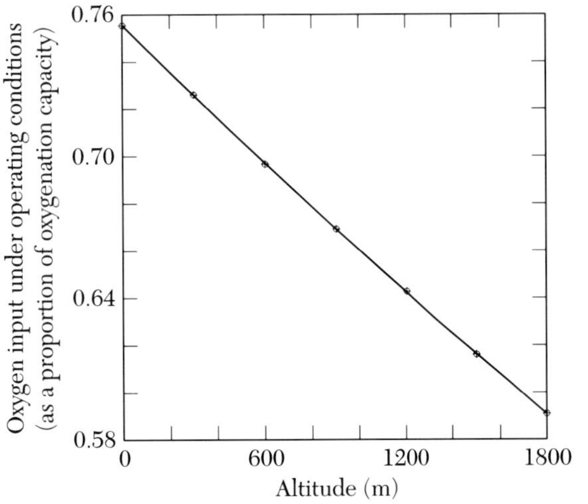

Figure 6.4. Oxygen input of aerators under operating conditions of 20 °C, 1 mg/l O_2 and α = 0.85.

is recommended because this ensures the minimum delay in returning sludge to the aeration tank, thereby reducing the likely occurrence of denitrification and the problem of rising sludge.

For small works, only one clarifier might be justified but where possible two should be provided as this will generally guarantee at least one tank in operation at all times.

(d) *Sludge return system.* The sludge return system includes the sludge scraper mechanism, sludge return sump, pumps and return pipeline.

Many types of pump are available, including centrifugal and screw pumps. The high cost of the latter might not be justified in a small works but they have the advantage of re-aerating the returned activated sludge. Standby pumping capacity is essential and multiple smaller units are generally better, giving greater operational flexibility than a single larger unit. It is recommended that the capacity of two pumps working together should be equal to the PWWF.

(ii) Types of oxidation ditch

The term oxidation ditch is used here in its widest sense, and covers all those activated sludge reactors that are designed as a continuous channel around which sewage is circulated under the action of various types of aerator.

(a) *Ditch configuration.* Oxidation ditch configuration is extremely variable, ranging from basic circular or oval shapes to more complex horseshoe or multiple-leg designs, the ultimate choice being influenced by factors such as flow, aeration requirements, topography and ground conditions. Similarly, the sides of the ditch can be either vertical or sloped. The depth ranges from 1.5 to 4.5 m and the design velocity in the ditch varies from 0.5 to 1.5 m/s.

Control of the velocity at the bottom of the ditch is necessary to prevent settlement of organic matter. This can be achieved by installing an inclined baffle downstream of the rotor, thereby deflecting the surface velocity downwards. It is suggested that a baffle arrangement should be used on ditches deeper than 2.0 m.

(b) *Inlet and outlet arrangements.* Because of the plug flow characteristics of oxidation ditches, the incoming sewage is not diluted as rapidly as in completely mixed systems. For domestic wastewater the feed is normally discharges at a single point into the aeration zone, although there are indications that feeding upstream of an aerator where DO is low can produce a sludge with better settling characteristics.

The position of the outlet is one of engineering convenience, although it is recommended that the outlet be located upstream of the inlet, with at least one aerator in the intervening flow path to minimize the likelihood of short circuiting.

(c) *Aerator performance.* Because aerators in oxidation ditches are used to provide kinetic energy to the water in addition to entraining air, the oxygen input is controlled by varying the depth of submersion of the aerator rather than by on/off speed control. Typical depths of submersion are 100–300 mm, providing an oxygenation variation of approximately 2:1. Typical oxygenation capacities for a 1 m diameter rotor are 8–12 kg O_2/h per metre of rotor length at design submergence.

The oxygen transfer efficiency is similar to that of low-speed vertical shaft aerators over the normal immersion depth operating range, at between 1.7 and 2.1 kg O_2/kW h. However, the figure does vary with depth and can reduce rapidly if the immersion depth is too great. This can place an excessive load on drive motors and provision should be made to limit the maximum immersion to prevent motor overload.

(d) *Loading rates.* Experience has indicated that these plants are adequately designed if they operate at very low loading rates, with sludge ages in excess of 15 days. This usually gives an hydraulic retention time of more than 30 h.

(e) *Orbal system.* This system is a specific configuration. It comprises concentric channels (usually three to five) with effluent moving from one to the next. A relatively long plug flow is thus obtained, providing for consecutive aerobic/anoxic zones with concomitant denitrification and partial removal of phosphate. Aeration/circulation is effected by rotating discs supported on a horizontal shaft.

The orbal system is particularly suitable for medium to smaller plants where minimum skilled supervision is available.

6.2.4 Design parameters for pond systems

Nuisance-free operation of a pond system is governed by the size and loading of the primary pond system, whereas effluent quality is determined by size and arrangement of the secondary ponds. The design parameters given below are the result of research done and practical experience gained in the Republic of South Africa, but these might have to be modified to suit specific conditions encountered at any particular locality.

The criteria applied to each system are as follows.

6.2.4.1 Facultative-aerobic pond system

(i) Primary pond

- Loading rate 2500 persons/ha per day
  or 135 kg BOD/ha per day
  or 270–300 kg COD/ha per day
- depth 1.5 m but 1.2 m minimum
- retention ≥25 days based on ADWF without recirculation.

The system is applicable where sufficient suitable land is available.

Recirculation can be effectively used in cases of high holiday loadings. A recirculation rate of 1:1 to 1:2 should be used during such periods.

(ii) Secondary ponds

The total retention time based on ADWF for the secondary ponds should be 25 days, with the first secondary pond having 10 days retention and the subsequent ponds 5 days each. These ponds should have a depth of 1.2–1.5 m.

6.2.4.2 Anaerobic–aerobic pond system

This system is suitable where the area of land is limited and well removed from residential areas, because anaerobic ponds produce noxious conditions and are unsightly.

(i) First anaerobic pond

- Loading rate 8 persons/m^3 per day
  or 0.4 kg BOD/m^3 per day
  or 0.85 kg BOD/m^3 per day
- depth 3–4 m
- retention ≥12 h at PDWF.

Recirculation from the primary pond or the first of the secondary ponds to the influent should be at a rate of 1:1 based on ADWF and should be applied at the surface to reduce odours.

(ii) Second anaerobic pond

The second anaerobic pond should be equal in size to the first pond. It collects suspended solids carried over from the first pond.

(iii) Primary aerobic pond

The retention in the primary aerobic pond should not be less than 15 days based on ADWF and the loading not more than 135 kg BOD/ha per day. The pond depth should be 1.2–1.5 m. The inlet to this pond should be arranged to ensure maximum dispersion.

(iv) Secondary aerobic ponds

The total retention time based on ADWF allowed for the secondary ponds is 25 days, with the first secondary having 10 days retention and the subsequent ponds 5 days each. These ponds should have a depth of 1.2–1.5 m.

6.2.4.3 Aerated lagoons

The aerated lagoon system is similar to the facultative-aerobic pond system except that the oxygen supply is augmented by surface aerators or diffusers, which allow the process to be intensified.

Because oxygen can thus be supplied throughout the day and night, reactions are faster and the retention time therefore shorter so that primary lagoons can be made deeper and smaller, thus also saving space.

Because aeration will give reasonably good mixing, a definite boundary between aerobic and anaerobic zones will hardly be evident.

The aerated lagoon is a much more homogeneous and stable reactor than the other pond types and is considerably more efficient, hence a loading rate of 600–900 kg BOD/ha per day can be applied on a 3 m deep pond. The power requirements to operate the aeration system amount to 10.8 kW h per 100 persons per day.

(i) Primary aerated pond

- Loading rate 20–30 g BOD/m^3 per day
- depth 2–3 m
- retention ≥5 days at ADWF
- oxygenation capacity 1.0–1.5 g O_2/g BOD.

(ii) Secondary aerated ponds

The secondary pond system should be designed on the basis set out in §6.2.4.1(ii).

6.2.4.4 Oxidation pond for night-soil

- Loading rate 3000–4000 persons/ha per day
  or 150–320 kg BOD/ha.d
- depth 2 m or deeper.

These systems are not designed on a volumetric basis.

The low water content of night-soil can be less than the amount lost by seepage and evaporation, so the pond could dry out and become unsightly and odorous. It is therefore an essential feature of a pond treating night-soil that make-up water be available to keep the pond filled. For ponds treating night-soil and conservancy tank contents there may be an overflow; therefore consideration should be given to the construction of an overflow pond.

6.2.4.5 Maturation ponds

Maturation ponds used as a final treatment stage in conjunction with biological filters or activated sludge plants should be designed on the same basis as the secondary pond system described in §6.2.4.1(ii) with retention times reduced to a total of 10 days, i.e. 4, 2, 2 and 2 days respectively in a series of four ponds. The pond depth should be between 1.2 and 1.5 m.

Index